MICROBIOLOGY RESEARCH ADVANCES

A CLOSER LOOK AT ANTIBIOTIC RESISTANCE

MICROBIOLOGY RESEARCH ADVANCES

Additional books and e-books in this series can be found on Nova's website under the Series tab.

MICROBIOLOGY RESEARCH ADVANCES

A CLOSER LOOK AT ANTIBIOTIC RESISTANCE

VICTOR TSUNODA
EDITOR

New York

NOTICE TO THE READER

The Publisher has taken reasonable care in the preparation of this book, but makes no expressed or implied warranty of any kind and assumes no responsibility for any errors or omissions. No liability is assumed for incidental or consequential damages in connection with or arising out of information contained in this book. The Publisher shall not be liable for any special, consequential, or exemplary damages resulting, in whole or in part, from the readers' use of, or reliance upon, this material. Any parts of this book based on government reports are so indicated and copyright is claimed for those parts to the extent applicable to compilations of such works.

Independent verification should be sought for any data, advice or recommendations contained in this book. In addition, no responsibility is assumed by the Publisher for any injury and/or damage to persons or property arising from any methods, products, instructions, ideas or otherwise contained in this publication.

This publication is designed to provide accurate and authoritative information with regard to the subject matter covered herein. It is sold with the clear understanding that the Publisher is not engaged in rendering legal or any other professional services. If legal or any other expert assistance is required, the services of a competent person should be sought. FROM A DECLARATION OF PARTICIPANTS JOINTLY ADOPTED BY A COMMITTEE OF THE AMERICAN BAR ASSOCIATION AND A COMMITTEE OF PUBLISHERS.

Additional color graphics may be available in the e-book version of this book.

Library of Congress Cataloging-in-Publication Data

ISBN: 978-1-53619-277-3

Published by Nova Science Publishers, Inc. † New York

CONTENTS

Preface		**vii**
Chapter 1	Antibiotic Resistance: Additional Federal Actions Needed to Better Determine Magnitude and Reduce Impact *United States Government Accountability Office*	**1**
Chapter 2	Combating Superbugs: U.S. Public Health Responses to Antibiotic Resistance *Committee on Energy and Commerce*	**155**
Index		**261**

PREFACE

Bacterial infections have become more difficult, and sometimes impossible, to treat due to antibiotic resistance, which occurs when bacteria develop the ability to defeat the available drugs designed to kill them. According to the Centers for Disease Control and Prevention, each year, 2 million Americans become sick with antibiotic-resistant infections, and of that, about 23,000 die. This book examines the challenges related to antibiotic resistance, the development and use of diagnostic testing to identify antibiotic resistance, the development of treatments for resistant infections, and appropriate antibiotic use.

Chapter 1 - Bacterial infections have become more difficult, and sometimes impossible, to treat due to antibiotic resistance, which occurs when bacteria develop the ability to defeat the available drugs designed to kill them. Concerns about rising rates of resistance to available treatment options prompted the federal government to create the 5-year National Action Plan in 2015. The plan called for federal agencies to strengthen surveillance, advance the development of diagnostic tests and new antibiotics, and slow the emergence of resistant bacteria, among other things.

GAO was asked to review federal efforts to address antibiotic resistance. This report examines federal efforts and challenges related to (1) surveillance of antibiotic resistance, (2) the development and use of diagnostic testing to identify antibiotic resistance, (3) the development of treatments for resistant infections, and (4) appropriate antibiotic use. GAO reviewed literature and agency documents; interviewed agency officials and health care industry, drug industry, and other stakeholders; and held a meeting of international and U.S. experts to obtain their views.

Chapter 2 - This is an edited, reformatted and augmented version of Hearing before the Subcommittee on Oversight and Investigations, of the Committee on Energy and Commerce, House of Representatives, One Hundred Fourteenth Congress, Second Session, Serial No. 114–153, dated June 14, 2016.

In: A Closer Look at Antibiotic Resistance ISBN: 978-1-53619-277-3
Editor: Victor Tsunoda

Chapter 1

ANTIBIOTIC RESISTANCE: ADDITIONAL FEDERAL ACTIONS NEEDED TO BETTER DETERMINE MAGNITUDE AND REDUCE IMPACT*

United States Government Accountability Office

ABBREVIATIONS

AHRQ	Agency for Healthcare Research and Quality
AR Option	Antimicrobial Resistance Option
AU Option	Antimicrobial Use Option
AUR Module	Antimicrobial Use and Resistance Module
BARDA	Biomedical Advanced Research and Development Authority
CARB	Combating Antibiotic-Resistant Bacteria

* This is an edited, reformatted and augmented version of United States Government Accountability Office, Report to Congressional Requesters, Accessible Version, Publication No. GAO-20-341, dated March 2020.

CARB-X	Combating Antibiotic-Resistant Bacteria Biopharmaceutical Accelerator
CDC	Centers for Disease Control and Prevention
CLSI	Clinical and Laboratory Standards Institute
CMS	Centers for Medicare & Medicaid Services
DOD	Department of Defense
eGISP	Enhanced Gonococcal Isolate Surveillance Project
EIP	Emerging Infections Program
FDA	Food and Drug Administration
GISP	Gonococcal Isolate Surveillance Program
GLASS	Global Antimicrobial Resistance Surveillance System
GPRA	Government Performance and Results Act
HHS	Department of Health and Human Services
LPAD	Limited Population Pathway for Antibacterial and Antifungal Drugs
MIPS	Merit-based Incentive Payment System
MRSA	Methicillin-resistant Staphylococcus aureus
NARMS	National Antimicrobial Resistance Monitoring System
NHSN	National Healthcare Safety Network
NIAID	National Institute of Allergy and Infectious Diseases
NIH	National Institutes of Health
PACCARB	Presidential Advisory Council on Combating Antibiotic- Resistant Bacteria
QIDP	Qualified Infectious Disease Product
R&D	research and development
SURRG	Strengthening the United States Response to Resistant Gonorrhea
TATFAR	Transatlantic Taskforce on Antimicrobial Resistance
VA	Department of Veterans Affairs
WHO	World Health Organization

Why GAO Did This Study

Bacterial infections have become more difficult, and sometimes impossible, to treat due to antibiotic resistance, which occurs when bacteria develop the ability to defeat the available drugs designed to kill them. Concerns about rising rates of resistance to available treatment options prompted the federal government to create the 5-year National Action Plan in 2015. The plan called for federal agencies to strengthen surveillance, advance the development of diagnostic tests and new antibiotics, and slow the emergence of resistant bacteria, among other things.

GAO was asked to review federal efforts to address antibiotic resistance. This report examines federal efforts and challenges related to (1) surveillance of antibiotic resistance, (2) the development and use of diagnostic testing to identify antibiotic resistance, (3) the development of treatments for resistant infections, and (4) appropriate antibiotic use. GAO reviewed literature and agency documents; interviewed agency officials and health care industry, drug industry, and other stakeholders; and held a meeting of international and U.S. experts to obtain their views.

What GAO Recommends

GAO is making eight recommendations to strengthen the federal response to combating antibiotic resistance. HHS concurred with seven recommendations and did not concur with one. More details are provided on the below.

In response to the findings presented in this Highlights, GAO recommends that:

- CDC ensure that its evaluation of its surveillance system for antibiotic-resistant gonorrhea includes measures of the system's representativeness of the U.S. population;

- HHS identify leadership and clarify roles and responsibilities to assess the clinical outcomes of diagnostic testing;
- HHS develop a strategy to further incentivize the development of new treatments for antibiotic- resistant infections, including through the use of postmarket financial incentives;
- HHS direct the CARB Task Force to include in its annual updates to the President plans for addressing any barriers preventing full implementation of the National Action Plan.

In addition, GAO is making four recommendations to address other CDC efforts in surveillance and reporting and to address FDA efforts in monitoring diagnostic tests.

HHS did not concur with the recommendation that it develop a strategy that includes the use of postmarket financial incentives to encourage the development of new treatments for antibiotic-resistant infections, citing its ongoing analysis to understand whether postmarket incentives should be included in such a strategy. GAO recognizes the complexity of this issue and maintains that this recommendation is warranted given that experts and others have called for additional postmarket incentives and the insufficiency of the current pipeline of new treatments for antibiotic-resistant infections.

WHAT GAO FOUND

The precise magnitude of the problem of antibiotic resistance is unknown. The Centers for Disease Control and Prevention (CDC) has made progress in expanding surveillance of infections from certain antibiotic-resistant bacteria in the United States and abroad but faces several challenges.

CDC faces challenges in conducting surveillance for antibiotic resistance due to the limited data it is able to collect through various surveillance systems. For example, CDC's primary surveillance system for gonorrhea—which CDC classified as an urgent antibiotic resistance threat affecting over half a million patients annually—currently tracks only an estimated 1 to 2 percent of all U.S. cases and only in males. CDC has not fully evaluated the representativeness of the gonorrhea surveillance system's results. However, it could do so, for example, by comparing the trends in their limited sample population with trends it can establish in the overall U.S. population via additional studies. Such an evaluation could give CDC more confidence that the system's data accurately reflect national trends.

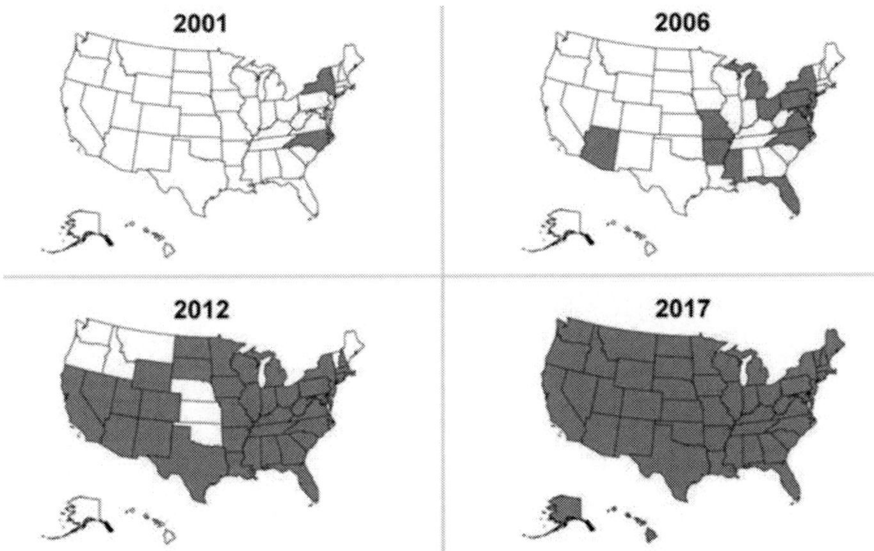

Note: This figure tracks a type of carbapenem-resistant Enterobacteriaceae (CRE), which, according to CDC, is a "nightmare bacteria" resistant to nearly all available antibiotics. Shading indicates CDC confirmed the presence of these bacteria within that state in that year or a previous one.

Source: Centers for Disease Control and Prevention (CDC). | GAO-20-341.

2001-2017 Cumulative Spread of One Type of Highly Resistant Bacteria in the United States.

Federal agencies have taken steps to advance the development and use of diagnostic tests to identify antibiotic-resistant bacterial infections, but these efforts have limitations. For example, agencies have conducted some studies to establish whether testing can lead to positive health care outcomes, such as reduced rates of antibiotic-resistant infections. However, more such studies are needed, according to experts and agency officials. Without information to guide test usage, clinicians may not be able to select appropriate treatments for their patients. One reason for the insufficient number of studies is that Department of Health and Human Services (HHS) agencies that are in a position to conduct or fund such studies—such as CDC and the Biomedical Advanced Research and Development Authority—disagree about what each agency should do. By clarifying roles and responsibilities, HHS agencies could more effectively address the need for more studies. The resulting studies could help demonstrate the value of diagnostic tests for antibiotic resistance, potentially increasing their use and improving patient care.

Experts warn that the current pipeline of antibiotics in development is insufficient to meet the threat of resistance. Several challenges impede the development of new treatments for resistant infections, notably inadequate return on investment for drug companies largely due to low prices and a limited patient population for whom these treatments would be appropriate. While HHS and Department of Defense agencies have provided financial premarket incentives to support antibiotic research and development, experts, federal officials and antibiotic developers agree that more postmarket incentives are needed to overcome the economic challenges. Advisory groups, including a presidential advisory council, and others have called for new postmarket incentives and identified multiple options for their design, including market entry rewards and reimbursement reform (see figure). However, HHS has not developed a strategy to further incentivize development of new treatments for antibiotic-resistant infections, and it may need to request authority and appropriations to create and implement certain types of incentives. Until such incentives are developed, more drug companies may exit the antibiotic development sector, and the pipeline of new treatments may continue to decrease.

Federal agencies have made several efforts to promote the appropriate use of antibiotics across health care settings through antibiotic stewardship—giving patients the right antibiotic at the right time, in the right dose, and for the right duration. However, key challenges remain. For example, federal agencies require only certain types of health care facilities to implement stewardship programs. In addition, CDC is limited in its ability to monitor and improve appropriate antibiotic use, in part because providers are not generally required to report antibiotic use data to a centralized database. The 5-year *National Action Plan for Combating Antibiotic-Resistant Bacteria* (National Action Plan) calls for strengthening antibiotic stewardship and for the timely reporting of antibiotic use data across health care settings. An executive order directs an interagency task force—the Combating Antibiotic-Resistant Bacteria (CARB) Task Force, coordinated by HHS—to provide annual updates to the President on, among other things, plans for addressing any barriers to full implementation of the National Action Plan. However, in its progress reports covering the first 4 years of the National Action Plan's implementation, the task force did not identify plans to address barriers to expanding antibiotic stewardship programs or the collection of antibiotic use data. Until it does so, the government will not have reasonable assurance that it is fully implementing the National Action Plan and addressing antibiotic resistance.

March 30, 2020

The Honorable Lamar Alexander
Chairman

The Honorable Patty Murray
Ranking Member
Committee on Health, Education, Labor, and Pensions
United States Senate

The Honorable Frank Pallone, Jr.
Chairman

The Honorable Greg Walden
Republican Leader
Committee on Energy and Commerce
House of Representatives

The Honorable Diana DeGette
Chair

The Honorable Brett Guthrie
Republican Leader
Subcommittee on Oversight and Investigations
Committee on Energy and Commerce
House of Representatives

Since the discovery of penicillin in 1928, many life-saving antibiotics have been developed that have allowed previously incurable infections to be easily treated. However, many types of infections have become more difficult or impossible to treat as bacteria have developed resistance to most—or, in some cases, all—currently available antibiotics. The Centers for Disease Control and Prevention (CDC) considers antibiotic resistance to be one of the greatest global public health threats of our time.[1] In 2019,

[1] Antimicrobial resistance refers broadly to drug-resistant bacterial, fungal, viral, and other types of microbial infections. For the purpose of this report, we focused on antibiotic resistance. CDC uses the term "antibiotics" to refer to drugs that can kill bacteria or fungi. See Department of Health and Human Services, Centers for Disease Control and Prevention, *Antibiotic Resistance Threats in the United States, 2019* (Atlanta, Ga.: Nov. 13, 2019), p. 7. Our review focused on resistance of bacteria against antibiotic drugs and did not focus on resistance of fungi against drugs. In contrast, under the Food, Drug and Cosmetic Act, the term "antibiotic drug" means "any drug (except drugs for use in animals other than humans) composed wholly or partly of any kind of penicillin, streptomycin, chlortetracycline, chloramphenicol, bacitracin, or any other drug intended for human use containing any quantity of any chemical substance which is produced by a microorganism and which has the capacity to inhibit or destroy microorganisms in dilute solution (including a chemically synthesized equivalent of any such substance) or any derivative thereof." 42 U.S.C. § 321(jj). Note that under the definition under the 21st Century Cures Act, which amended Section 319E of the Public Health Service Act, that for certain purposes of the law, the term "antimicrobial" includes "any antibacterial or antifungal drugs, and may include drugs that eliminate or inhibit the growth of other microorganisms, as appropriate." 42 U.S.C. § 247d-5(k).

CDC estimated that at least 2.8 million people get sick and at least 35,900 die each year from antibiotic-resistant infections in the United States.[2]

While bacteria naturally develop resistance to antibiotics over time, this problem has been accelerated by the overuse and misuse of antibiotics in human health, food animals, and the environment. The World Health Organization (WHO) has warned that the world urgently needs to change the way antibiotics are prescribed and used, and CDC has highlighted the need for antibiotics to be used more appropriately—a concept called antibiotic stewardship—to preserve their effectiveness and help slow the development of antibiotic resistance.[3] CDC officials noted that poor infection control and limited communication between health care facilities also contribute to the spread of antibiotic resistance. Furthermore, WHO and others warned that the pipeline of antibiotics in development is

---

[2] Centers for Disease Control and Prevention, *Antibiotic Resistance Threats, 2019.* In 2013, CDC had estimated at least 2 million people got sick and 23,000 people died each year from antibiotic-resistant infections; however, in its 2019 report, CDC revised its 2013 estimates by using new data sources and stated that the annual number of antibiotic- resistant infections in 2013 was at least 2.6 million, with 44,000 deaths. See Department of Health and Human Services, Centers for Disease Control and Prevention, *Antibiotic Resistance Threats in the United States, 2013* (Atlanta, Ga.: Apr. 23, 2013).

A study funded by the European Centre for Disease Prevention and Control estimated 671,689 antibiotic-resistant infections and 33,110 deaths caused by antibiotic-resistant infections in 2015 throughout the European Union and the European Economic Area. In addition, the study estimated the health burden of five types of infections with antibiotic-resistant bacteria using a metric called disability-adjusted life-years. A. Cassini et al., "Attributable Deaths and Disability-Adjusted Life-Years Caused By Infections with Antibiotic-Resistant Bacteria in the EU and the European Economic Area in 2015: a Population-Level Modelling Analysis," *Lancet Infectious Disease,* vol. 19, no. 1 (2019), pp. 56-66.

A 2014 study estimated that as many as 10 million people globally could die each year by 2050 if no action is taken to combat antimicrobial resistance. This study examined a limited number of types of bacteria, as well as other pathogens such as malaria, noting that this limitation resulted from a lack of readily available data. The Review on Antimicrobial Resistance, *Tackling a Crisis for the Health and Wealth of Nations.* (December 2014).

[3] World Health Organization, "Antibiotic Resistance," accessed February 13, 2020, https://www.who.int/en/news-room/fact-sheets/detail/antibiotic-resistance. WHO is the Geneva-based coordinating authority on international health within the United Nations system.

CDC defines antibiotic stewardship as the effort to measure and optimize antibiotic use with the goal of optimizing the treatment of infections while reducing the adverse events associated with antibiotic use. Antibiotic stewardship aims to have all patients treated with the right antibiotic at the right time, in the right dose, and for the right duration for a given diagnosis.

insufficient to tackle the growing threat of antibiotic resistance.[4] Additionally, diagnostic testing used to identify antibiotic-resistant bacteria is not available for all bacteria of concern. These gaps may hinder the correct diagnosis of antibiotic-resistant infections, which could delay treatment with appropriate antibiotics, contribute to antibiotic overuse, and impede overall surveillance efforts.[5]

Recognizing the growing threat of antibiotic resistance, by Executive Order No. 13676, September 2014, the President established the Task Force for Combating Antibiotic-Resistant Bacteria (CARB Task Force), co-chaired by the Secretaries of the Departments of Health and Human Services (HHS), Defense (DOD), and Agriculture.[6] In 2015, the CARB Task Force issued the *National Action Plan for Combating Antibiotic-Resistant Bacteria* (hereafter referred to as the National Action Plan), setting forth goals over 5 years to slow the development of resistant bacteria, strengthen national surveillance efforts, advance the development and use of diagnostic tests, and accelerate the development of new treatments, among other things.[7] Because of the severity of the problem that antibiotic resistance presents for humans and the federal government's commitment to fight it, you asked us to provide information on federal efforts to combat antibiotic resistance. This report examines

1) CDC's efforts to conduct surveillance of antibiotic resistance and any challenges to these efforts;
2) federal efforts to advance the development and use of tests for diagnosing antibiotic-resistant infections;
3) challenges to developing new treatments for antibiotic-resistant infections and federal efforts to address the challenges; and

---

[4] World Health Organization, *2019 Antibacterial Agents in Clinical Development: An Analysis of the Antibacterial Clinical Development Pipeline* (Geneva, Switzerland: 2019).

[5] Disease surveillance is the process of reporting, collecting, analyzing, and exchanging information related to cases of infectious diseases.

[6] Exec. Order No. 13676, *Combating Antibiotic-Resistant Bacteria*, 79 Fed. Reg. 56931 (Sept. 23, 2014).

[7] The White House, *National Action Plan for Combating Antibiotic-Resistant Bacteria* (Washington, D.C.: March 2015).

4) federal efforts to promote the appropriate use of antibiotics and any challenges that remain.

To address all four objectives, we reviewed relevant agency reports and documents; interviewed officials from federal agencies, experts on topics related to antibiotic resistance, and representatives from stakeholder organizations; reviewed relevant laws, regulations, policies, literature, and GAO reports; and attended two national conferences.[8] We focused our review primarily on agency actions since 2015, when the National Action Plan was published. We also focused our review on human health, as we have reported on federal efforts to address the use of antibiotics in food animals and recommended actions to improve these efforts for more than 20 years.[9] Additionally, we focused our review on antibiotic-resistant bacteria. Finally, we generally excluded federal efforts related to infection prevention and control in human health care, on which we have previously reported.[10]

We interviewed officials from federal agencies responsible for implementing the aspects of the National Action Plan related to our objectives: HHS's Office of the Assistant Secretary for Planning and Evaluation, the Biomedical Advanced Research and Development Authority (BARDA), CDC, the Centers for Medicare & Medicaid Services (CMS), the Food and Drug Administration (FDA), the National Institutes of Health (NIH), and the Office of Global Affairs; as well as DOD and the Department of Veterans Affairs (VA). We also interviewed experts and representatives from organizations involved in public health and

---

[8] The two conferences were the World Anti-Microbial Resistance Congress and the Gordon Research Conference on chemical and biological threat defense.

[9] GAO, *Antibiotic Resistance: More Information Needed to Oversee Use of Medically Important Drugs in Food Animals*, GAO-17-192 (Washington, D.C.: Mar. 2, 2017); *Antibiotic Resistance: Agencies Have Made Limited Progress Addressing Antibiotic Use in Animals*, GAO-11-801 (Washington, D.C.: Sept. 7, 2011); *Antibiotic Resistance: Federal Agencies Need to Better Focus Efforts to Address Risk to Humans from Antibiotic Use in Animals*, GAO-04-490 (Washington, D.C.: Apr. 22, 2004); and *Food Safety: The Agricultural Use of Antibiotics and Its Implications for Human Health*, GAO/RCED-99-74 (Washington, D.C.: Apr. 28, 1999).

[10] GAO, *Health-Care-Associated Infections in Hospitals: Leadership Needed from HHS to Prioritize Prevention Practices and Improve Data on These Infections*, GAO-08-283 (Washington, D.C.: Mar. 31, 2008).

epidemiology, infectious diseases and microbiology, antibiotic research and development (R&D), antibiotic stewardship, and other issues relating to antibiotic resistance. Because antibiotic resistance is a global problem, we also interviewed officials from WHO and other international entities. We identified experts and organizations through literature and other documents we reviewed and through referrals from agency officials and other experts we interviewed. In addition, in September 2018, with the assistance of the National Academy of Sciences, we convened a meeting of experts in antibiotic resistance epidemiology, diagnostic testing, antibiotic development, and antibiotic stewardship. (In this report, we refer to such experts as "experts at our meeting.") For each of our objectives, we identified and reported on actions taken by federal agencies to address antibiotic resistance, and we evaluated these actions against relevant criteria, as applicable, such as Standards for Internal Control in the Federal Government and GAO's leading practices on interagency collaboration.[11]

To examine CDC's efforts to conduct surveillance for antibiotic resistance and any challenges to these efforts, we reviewed documentation and conducted interviews with agency officials and other key stakeholders on each of the surveillance systems across CDC that collects antibiotic resistance data and reviewed CDC's *Antibiotic Resistance Threats in the United States, 2013* (2013 Threats Report) and *Antibiotic Resistance Threats in the United States, 2019* (2019 Threats Report) reports.[12] This included a review of health care facility participation data by state and territory in a CDC antibiotic resistance reporting system. We assessed the reliability of these data by reviewing them for any outliers or anomalies and inquiring with agency officials about their source and any known reliability issues. We determined that these data were sufficiently reliable for assessing facility participation rates by U.S. state and territory. We also reviewed documents from WHO's Global Antimicrobial Resistance

---

[11] GAO, *Standards for Internal Control in the Federal Government*, GAO-14-704G (Washington, D.C.: Sept. 10, 2014) and *Managing for Results: Key Considerations for Implementing Interagency Collaborative Mechanisms,* GAO-12-1022 (Washington, D.C.: Sept. 27, 2012).

[12] We started our audit work in 2018, prior to the issuance of CDC's 2019 Threats Report.

Surveillance System and interviewed WHO and CDC officials to identify challenges that limit CDC's ability to assess threats from abroad.

To examine federal efforts to advance the development and use of diagnostic tests, we also interviewed a nongeneralizable sample of six diagnostic test manufacturers to encompass different types of tests, based a list of manufacturers compiled from our previous work, interviews with select experts, and internet research. We limited our scope to FDA-authorized tests—which we are defining as tests that have been reviewed and cleared or granted authorization by FDA for marketing in the United States—that can identify resistance in at least one type of bacteria categorized as a priority in CDC's 2013 Threats Report.[13] Some of these tests are called antibiotic susceptibility tests, but we use "tests" to refer to the entire class of such tests. We included in our scope tests that can differentiate between viral and bacterial infections, because these types of tests are included in the National Action Plan.

To identify challenges to developing new treatments for antibiotic-resistant infections and examine federal efforts to address these challenges, we also reviewed literature and reports written by health policy advisory groups on topics related to antibiotic development challenges and incentives for development. Our examination of challenges and related federal actions focused on treatments and, therefore, did not include products designed to prevent infections, such as vaccines. We also interviewed 11 randomly selected companies that conduct R&D on new treatments for bacterial infections.[14]

To examine federal agency efforts to promote the appropriate use of antibiotics and any challenges that remain, we also analyzed CMS data and related documentation on quality measures and improvement activities

---

[13] FDA officials told us that the tests under our review are Class II devices and are appropriately referred to as cleared when granted marketing authorization. For a further description of device classes, see our previous report, GAO, *Medical Devices: Challenges and Capabilities to Enable Rapid Diagnoses of Infectious Diseases*, GAO-17-347 (Washington, D.C.: Aug. 14, 2017). Other types of diagnostic tests, such as laboratory- developed tests, are not within the scope of this report. FDA describes laboratory- developed tests as tests designed, manufactured, and used within a single laboratory.

[14] We selected from among companies that are researching or developing antibiotics and alternatives to antibiotics—which we call "nontraditional products" in this report—and we included companies that do and do not have existing FDA-approved drugs on the market.

related to antibiotics. We reviewed the data for any obvious outliers or anomalies and determined that these data were sufficiently reliable for reporting on the number of clinicians who reported implementing these quality measures and improvement activities. In addition, we reviewed aggregated data from CDC on the total number of eligible U.S. hospitals reporting their antibiotic use data to a CDC system. We assessed the reliability of the aggregated data by reviewing them for any obvious errors or missing data totals and inquiring with agency officials about their source and any known reliability issues. We determined that these data were sufficiently reliable for reporting hospital participation rates in CDC's antibiotic use reporting system. We focused our review on antibiotic use in the United States, rather than global antibiotic use.

Appendix I contains more detailed information on the scope and methodology of our review. Appendixes I and II contain more detailed information about our expert meeting.

We conducted this performance audit from February 2018 to March 2020 in accordance with generally accepted government auditing standards.

Those standards require that we plan and perform the audit to obtain sufficient, appropriate evidence to provide a reasonable basis for our findings and conclusions based on our audit objectives. We believe that the evidence obtained provides a reasonable basis for our findings and conclusions based on our audit objectives.

BACKGROUND

Antibiotics are drugs that work by killing bacteria or slowing their growth.[15] However, some bacteria have developed ways to resist the effects of antibiotics, for example, by preventing antibiotics from entering

---

[15] We note that in its 2019 report, consistent with applicable authorities, CDC uses the term "antibiotics" to refer to drugs that can kill bacteria or fungi. CDC's *Antibiotic Resistance Threats, 2019*, p. 7. See also, 21 U.S.C. § 321(jj); 42 U.S.C. § 247d-5(k). However, as we explained previously, we have only considered those antibiotics that can kill bacteria.

the cell or pumping them out after the antibiotic enters. Bacteria that are able to survive in the presence of antibiotics will multiply and pass on their new genetic material that confers resistance to future generations of bacteria and, in some cases, to other types of bacteria.[16] Resistance can arise in bacteria in humans, animals, and the environment, including in health care settings, and can spread through contact with infected people or animals, contact with contaminated water, soil or surfaces, or consumption of contaminated food.[17]

The spread of antibiotic resistance threatens not only the ability to fight bacterial infections but also threatens to reverse some significant medical gains. For example, in addition to treating infections, antibiotics have allowed for numerous medical procedures, such as joint replacements, caesarian sections, organ transplants, chemotherapy, and dialysis—all of which would be significantly riskier without effective antibiotics. Antibiotic resistance also poses a significant economic burden resulting from the direct costs of treating those with resistant infections and the loss of economic productivity from those who get sick or die.[18]

In the 2013 Threats Report, CDC identified 17 bacterial pathogens that the agency considers to be "urgent," "serious," or "concerning" because they have developed enough resistance to antibiotics to be considered a threat to human health. (See Figure 1.) According to CDC, certain types of bacteria, called gram-negative bacteria, are particularly worrisome because they are becoming resistant to nearly all drugs that would be considered for treatment.[19]

---

[16] Scientists have identified more than 1,600 genes that confer resistance for bacteria. Of significant global scientific concern is the *mcr* gene, which confers resistance to polymixins, known as the "drugs of last resort" for treating certain types of infections. The *mcr* gene was first discovered in China in 2015 but has since been found in bacteria cultured from humans and animals on at least five continents, including in the United States.

[17] The National Action Plan includes a goal to strengthen surveillance of antibiotic resistance using a "One-Health" approach, which recognizes the interaction between humans, animals, and the environment.

[18] According to CDC, antibiotic-resistant infections require extended hospital stays, additional follow-up visits to health care providers, and the use of treatments that may be more costly and potentially more toxic.

[19] Gram-negative bacteria are characterized by a double cell membrane, which can pose as a barrier to antibiotics trying to enter the cell, and pumps that can expel antibiotics from the cell.

The most serious gram-negative infections can be acquired in hospitals or other health care settings and can cause pneumonia, bloodstream infections, wound or surgical site infections, and meningitis. Nine of the 17 bacterial threats on CDC's threat list are gram-negative. One of the bacteria CDC considers to be an urgent threat—*Clostridioides difficile* (*C. difficile*)—is classified as a threat not because it is resistant to antibiotics, but because it is caused by the same factors that drive antibiotic resistance, such as antibiotic use. CDC estimates that *C. difficile* alone accounted for 12,800 deaths in U.S. hospitals in 2017.[20] CDC's 2013 Threats Report also identified one type of fungus—*Candida auris*—that it considered to be a serious threat (see text box).

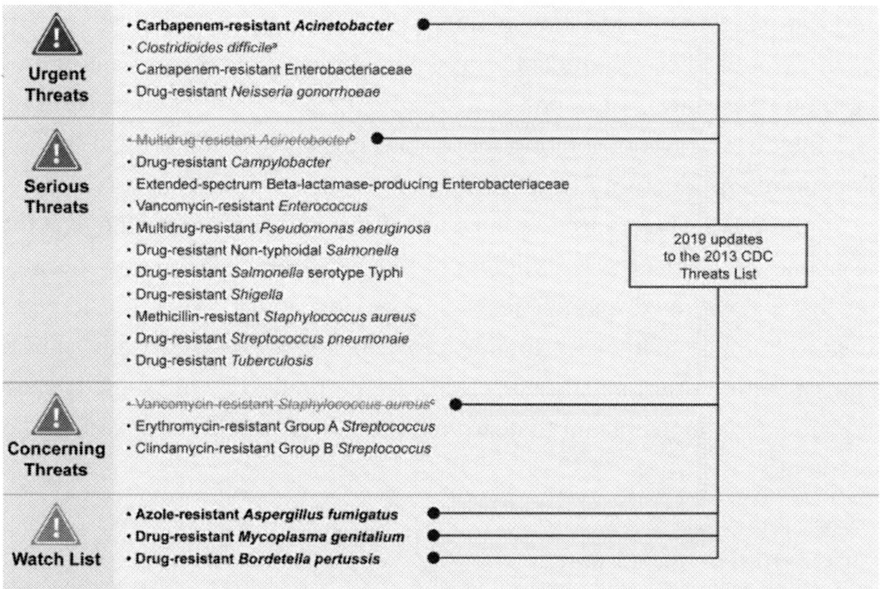

Source: Centers for Disease Control and Prevention (CDC). | GAO-20-341.

[20] The deaths caused by *C. difficile* are not included in the 35,000 deaths that CDC attributed to antibiotic-resistant infections in CDC's *Antibiotic Resistance Threats, 2019*. *Clostridioides difficile* was formerly known as *Clostridium difficile*.

Data table for Figure 1. Bacteria CDC considered to be threats in 2013 and 2019

Threat level	Bacteria
Urgent Threats	Carbapenem-resistant Acinetobacter (2019 update to the 2013 CDC Threats List) *Clostridioides difficile*[a] Carbapenem-resistant Enterobacteriaceae Drug-resistant *Neisseria gonorrhoeae*
Serious Threats	Multidrug-resistant *Acinetobacter*[b] (removed from serious threat level in 2019 update to the 2013 CDC Threats List) Drug-resistant *Campylobacter* Extended-spectrum Beta-lactamase-producing Enterobacteriaceae Vancomycin-resistant *Enterococcus* Multidrug-resistant *Pseudomonas aeruginosa* Drug-resistant Non-typhoidal *Salmonella* Drug-resistant *Salmonella* serotype Typhi Drug-resistant *Shigella* Methicillin-resistant S*taphylococcus aureus* Drug-resistant *Streptococcus pneumonaie* Drug-resistant *Tuberculosis*
Concerning Threats	Vancomycin-resistant *Staphylococcus aureus*[c] (removed from concerning threat level in 2019 update to the 2013 CDC Threats List) Erythromycin-resistant Group A *Streptococcus* Clindamycin-resistant Group B *Streptococcus*
Watch List (2019 updates to the 3013 CDC Threats List)	Azole-resistant *Aspergillus fumigatus* Drug-resistant *Mycoplasma genitalium* Drug-resistant *Bordetella perstussis*

Note: In 2013, CDC also considered fluconazole-resistant *Candida*, which is a fungus, to be a serious threat. In 2019, CDC added *Candida auris* as an urgent threat and focused on drug-resistant *Candida* as a serious threat.

[a]*Clostridioides difficile* was formerly known as *Clostridium difficile*.

[b]In 2019, CDC shifted *Acinetobacter* from its list of serious threats to its list of urgent threats because of the emergence of easily spread resistance and the lack of antibiotics currently available or in development to treat these infections. CDC also reported that, in 2019, they focused on "carbapenem-resistant" *Acinetobacter*, instead of "multidrug-resistant" *Acinetobacter*.

[c]CDC dropped Vancomycin-resistant *Staphylococcus aureus* from its threats list in 2019, noting that the 14 identified cases of infection with this bacteria were considered isolated cases, and spread from patient to patient has never been documented.

U.S. spending on antibiotics in health care from 2010 through 2015 was estimated in one study to be nearly $56 billion, ranging from $8.4 billion to $10.6 billion annually.[21]

[21] Spending estimates were based on an analysis of data on purchases by health care settings and public services in the United States. See K. J. Suda et al., "Antibiotic Expenditures by

Candida auris Is a Resistant Fungal Threat

Candida auris (*C. auris*) is an emerging infectious fungus that, according to the Centers for Disease Control and Prevention (CDC), presents a global health threat in part because it is highly resistant to anti-fungal drugs and is challenging to address. *C. auris* was first identified in Japan in 2009. CDC reported 806 confirmed cases in the United States, as of August 31, 2019. According to CDC, *C. auris* is highly transmissible and some commonly used hospital surface disinfectants appear to be less effective against *C. auris*. A CDC official told us *C. auris* is a good example of an emerging threat that requires more research and associated efforts to properly address.

Addressing *C. auris* is challenging for reasons including the rise of resistance and limitations in diagnostic tests. According to CDC, there are three classes of antifungals available to treat *C. auris*. However, CDC has identified strains that are resistant to all three classes. A CDC official noted that getting new antifungals to market is challenging because, among other things, the demand for antifungals, relative to antibiotics, is low. Additionally, according to FDA, although reliable tests for identifying *C. auris* exist, commonly used laboratory tests may misidentify this fungus, posing a barrier to correct diagnosis. In 2018, the Food and Drug Administration (FDA) cleared a test based on mass spectrometry to identify *C. auris*, but this test cannot characterize resistance. FDA officials told us there are three FDA-cleared tests available for testing for other Candida species' resistance to fluconazole. However, none of these tests can provide rapid results, such as within an hour. Finally, interpretation of culture-based diagnostic tests, which examine how well bacteria grow in the presence of an antibiotic, is challenging due to the lack of established interpretive criteria for *C. auris*, by both the Clinical and Laboratory Standards Institute, which promotes the development and use of voluntary laboratory consensus standards and guidelines within the health care community, and by FDA.

Source: GAO summary of CDC interviews and website and FDA interviews. | GAO-20-341.

While CDC states that antibiotic prescribing improved nationally with a 5 percent decrease from 2011 to 2016, the agency estimated in 2017 that at least 30 percent of antibiotics used across both outpatient and inpatient settings are still prescribed unnecessarily or incorrectly and, therefore, are considered inappropriate.[22]

Medication, Class, and Healthcare Setting in the United States, 2010- 2015," *Clinical Infectious Diseases*, vol. 66, no. 2 (2018): pp. 185-90.

[22] Department of Health and Human Services, Centers for Disease Control and Prevention, *Antibiotic Use in the United States, 2017: Progress and Opportunities* (Atlanta, Ga.: 2017)

Antibiotic Use in the United States

A 2017 Centers for Disease Control and Prevention (CDC) report estimates that about 30 percent of antibiotics used in U.S. hospitals are inappropriate (unnecessary or prescribed incorrectly), and as much as 50 percent of antibiotics prescribed in outpatient settings—such as physicians' offices, emergency departments, urgent care centers, and retail clinics—may be inappropriate. For example, CDC reports that each year, an estimated 47 million unnecessary antibiotic prescriptions are written in physicians' offices and emergency departments. Most of these unnecessary prescriptions are for respiratory conditions most commonly caused by viruses— including common colds, viral sore throats, and bronchitis—that do not respond to antibiotics, or for bacterial infections that do not always need antibiotics, like many sinus and ear infections. Furthermore, CDC reports that even when antibiotics are needed, prescribers often favor drugs that may be less effective and may carry more risk over more targeted, "first-line" drugs recommended by nationally recognized antibiotic prescribing guidelines. (First-line drugs are the drugs generally recommended for initial treatment for a given diagnosis, often combining the best efficacy with the best safety profile or the lowest cost.) According to CDC, antibiotics are among the most frequently prescribed medications in nursing homes, with up to 70 percent of residents receiving one or more courses of systemic (non-topical) antibiotics in a year; CDC also cites studies showing that 40 to 75 percent of antibiotics prescribed in nursing homes may be inappropriate. CDC further reports that harms from antibiotic overuse include the risk of serious diarrheal infections from *C. difficile*, increased adverse drug events and drug interactions, and increased risk of infection with antibiotic-resistant organisms.

Source: CDC, Antibiotic Use in the United States, 2017: Progress and Opportunities (Atlanta, Ga.: 2017), The Core Elements of Antibiotic Stewardship for Nursing Homes (Atlanta, Ga.: 2015), and "Measuring Outpatient Antibiotic Prescribing," accessed August 27, 2019, https://www.cdc.gov/antibiotic-use/community/programs-measurement/measuring-antibiotic-prescribing.html. | GAO-20-341.

According to CDC, approximately 85 to 95 percent of the nation's antibiotic use, by volume, occurred in outpatient settings from 2010 through 2015; and roughly 270 million antibiotic prescriptions—

---

and *Antibiotic Use in the United States, 2018 Update: Progress and Opportunities* (Atlanta, Ga.: 2019).

According to CDC officials, "unnecessary" antibiotic use means the antibiotic was prescribed when no antibiotic was needed, based on clinical practice guidelines. "Inappropriate" antibiotic use includes both unnecessary antibiotic use, as well as inappropriate antibiotic selection, dosing, or duration when antibiotics are indicated. CDC officials also told us they consider "misuse" and "inappropriate use" to be synonymous terms.

equivalent to 836 per 1,000 persons in the United States—were written in these settings in 2016. (For more information on antibiotic use in the United States, see text box.)

The National Action Plan and Federal Agency Responsibilities

In September 2014, the President signed Executive Order No. 13676 (Executive Order), which directed that several federal actions be initiated related to antibiotic resistance. For example, the Executive Order directed the creation of the National Action Plan, which the White House released in 2015, to provide a roadmap for federal agencies to respond to the threat of antibiotic resistance.[23] The National Action Plan set five major goals over 5 years related to (1) slowing the emergence of resistant bacteria and preventing the spread of resistant infections; (2) strengthening national One-Health surveillance efforts to combat resistance;[24] (3) advancing the development and use of rapid and innovative diagnostic tests for the identification and characterization of resistant bacteria; (4) accelerating basic and applied R&D for new antibiotics, other therapeutics, and vaccines; and (5) improving international collaboration and capacities related to the first four goals.[25] In addition, the National Action Plan discusses the importance of preventing and controlling infections, such as through rapid detection, to combat antibiotic resistance domestically and globally (see text box). Within each of these five goals, the National Action Plan contains numerous objectives, sub-objectives, agency-specific milestones, and other performance targets called significant outcomes. For

---

[23] Exec. Order No. 13676 (Sept. 23, 2014).

[24] One-Health recognizes the interaction between humans, animals, and the environment.

[25] Similarly, the objectives of WHO's *Global Action Plan on Antimicrobial Resistance* are to (1) improve awareness and understanding of antimicrobial resistance though effective communication, education, and training; (2) strengthen the knowledge and evidence base through surveillance and research; (3) reduce the incidence of infection through effective sanitation, hygiene, and infection prevention measures; (4) optimize the use of antimicrobial medicines in human and animal health; and (5) develop the economic case for sustainable investment that takes account of the needs of all countries and to increase investment in new medicines, diagnostic tools, vaccines and other interventions. World Health Organization, *Global Action Plan on Antimicrobial Resistance* (Geneva, Switzerland: 2015).

example, the National Action Plan set a significant outcome of reducing inappropriate antibiotic use by 50 percent in outpatient settings and by 20 percent in inpatient settings by 2020.

The interagency CARB Task Force, which was created by the Executive Order to issue and monitor the implementation of the National Action Plan, is co-chaired by the Secretaries of Defense, Agriculture, and HHS, and is additionally comprised of representatives from VA and several other agencies. Representatives from HHS agencies—including BARDA, CDC, CMS, FDA, and NIH—make up nearly two-thirds of the task force's participants (see Table 1).

Vaccines Can Also Help Prevent Antibiotic Resistance

While we did not include vaccines in the scope of this report, vaccines play a role in helping combat antibiotic resistance because they are designed to prevent infections, including resistant infections. In addition, by preventing infections from occurring, they can reduce the need to use antibiotics, which in turn, can slow the development of antibiotic resistance. For example, according to the Centers for Disease Control and Prevention (CDC), since introduction of the pneumococcal conjugate vaccine among children in 2000, rates of antibiotic-resistant infections caused by certain *Streptococcus pneumoniae* strains decreased by 97 percent among children under 5 and by more than 60 percent among adults. However, few vaccines are available that target antibiotic-resistant bacteria on CDC's threat list.

Source: GAO summary of CDC information. | GAO-20-341.

According to the HHS Assistant Secretary for Planning and Evaluation officials who coordinate it, the task force is developing a new National Action Plan that will span the years 2020 through 2025. To provide additional advice to the CARB Task Force and the Secretary of HHS, the Executive Order also created the Presidential Advisory Council on Combating Antibiotic-Resistant Bacteria (PACCARB), which is composed of 15 non-governmental members.[26]

[26] Exec. Order No. 13676, § 4 Presidential Advisory Council on Combating Antibiotic- Resistant Bacteria, 79 Fed. Reg. 56931 (Sept. 23, 2014); continued by Exec. Order No. 13708 § 1(jj), 80 Fed. Reg. 60271 (Sept. 30, 2015), and Exec. Order No. 13811 § 1(z), 82 Fed. Reg. 46363 (Sept. 29, 2017). In June 2019, Congress passed the Pandemic and All- Hazards

Infection Prevention and Control

According to the World Health Organization (WHO), effective infection prevention and control measures are a practical and scientific approach to reduce health care- associated infections in patients and health care workers, and help combat antibiotic resistance. Infection prevention and control measures serve as the cornerstone of actions needed to address epidemics, pandemics, and antibiotic resistance. Such measures include implementing hand hygiene practices, providing vaccinations, cleaning and disinfecting hospital rooms, isolating patients with infectious diseases, decontaminating and sterilizing medical equipment, and tracking data about emerging infectious diseases. WHO states that health care-associated infections are a global challenge from which no country or health care facility is immune. The Centers for Disease Control and Prevention (CDC) has taken actions to address and track health care-associated infections, including antibiotic-resistant infections. For example, in 2009, CDC issued guidance for infection control targeting Enterobacteriaceae that may be resistant to carbapenem, a class of antibiotics. In 2018, CDC published a study suggesting that a tracked decline in the proportion of resistant bacteria, including carbapenem-resistant Enterobacteriaceae, observed in some health care settings, could be attributable—at least in part—to actions such as those outlined in its 2009 guidance. In addition, CDC has reported that U.S. hospitals have made major progress since 2005 in declining rates of methicillin-resistant *Staphylococcus aureus* (MRSA) bacteremia because of infection prevention measures.

Source: GAO summary of WHO and CDC documents. | GAO-20-341.

The Executive Order also charged the CARB Task Force with providing annual updates to the President regarding progress made in implementing the National Action Plan, plans to address any barriers preventing its full implementation, and recommendations for any new or modified actions, taking federal government resources into consideration. Since 2015, the CARB Task Force has produced four progress reports, which summarize agency actions toward meeting the goals and milestones

---

Preparedness and Advancing Innovation Act of 2019, which transitioned authority for the PACCARB from Executive Order to law. Pandemic and All-Hazards Preparedness and Advancing Innovation Act of 2019, Pub. L. No. 116-22, § 505, 133 Stat. 905 (June 24, 2019), 42 U.S.C. § 247d–5, note.

PACCARB has produced four reports, including one published in July 2019 with recommendations for the next National Action Plan.

laid out in the National Action Plan; these reports were provided to the President and are publicly available.[27]

Table 1. Examples of key roles and responsibilities of select Department of Health and Human Services Agencies related to combating antibiotic resistance

Agency	Examples of key roles and responsibilities
Agency for Healthcare Research and Quality (AHRQ)	AHRQ's mission is to produce evidence to make health care safer, higher quality, more accessible, equitable, and affordable. AHRQ funds research to develop improved methods for combating antibiotic resistance and conducting antibiotic stewardship. AHRQ also conducts nationwide projects that work with front-line clinicians to promote the implementation of evidence-based methods for infection prevention and antibiotic stewardship in hospitals, long-term care, and ambulatory care settings.
Biomedical Advanced Research and Development Authority (BARDA)	BARDA's mission is to help secure the nation from chemical, biological, radiological, and nuclear threats, as well as from pandemic influenza and emerging infectious diseases. Through public-private partnerships, BARDA invests in products, such as antibiotics and diagnostic tests, which can be used to fight the threat of antibiotic resistance. BARDA, along with the National Institutes of Health, helped launch the Combating Antibiotic-Resistant Bacteria Biopharmaceutical Accelerator (CARB-X) in 2016, which is a public-private international partnership that funds research and development of new antibiotics, diagnostic tests, vaccines, and other treatments to combat antibiotic resistance.[a]
Centers for Disease Control and Prevention (CDC)	CDC has a primary responsibility to protect the public health through the prevention of disease and health promotion. CDC monitors the emergence of antibiotic-resistant infections and the use of antibiotics. CDC provides state and local health departments with resources to detect and track antibiotic-resistant pathogens, support antibiotic-resistance experts to implement infection control activities, and respond to outbreaks to stop the spread, in addition to promoting appropriate use of antibiotics in human health care to prevent the emergence of new resistance. CDC also supports research on and implements prevention strategies, such as educational programs targeting the public and health care providers that are designed to provide information on the appropriate use of antibiotics.

[27] United States Taskforce for Combating Antibiotic-Resistant Bacteria (CARB Task Force), *National Action Plan for Combating Antibiotic-Resistant Bacteria: First 180 Days Report* (November 2015), *National Action Plan for Combating Antibiotic-Resistant Bacteria: Progress Report for Years 1 and 2* (October 2017), *National Action Plan for Combating Antibiotic-Resistant Bacteria: Progress Report: Year 3* (October 2018), and *National Action Plan for Combating Antibiotic-Resistant Bacteria: Progress Report: Year 4* (September 2019).

Table 1. (Continued)

Agency	Examples of key roles and responsibilities
Centers for Medicare & Medicaid Services (CMS)	CMS is responsible for administering the Medicare and Medicaid programs. Medicare provides federally financed health insurance coverage to people age 65 and older and certain other individuals. Medicaid is a joint federal-state program that finances health care coverage for low-income and medically needy individuals. CMS sets payment rates for Medicare beneficiaries receiving treatment in hospitals, including those with antibiotic-resistant infections. In addition, CMS determines patient health and safety requirements for certain types of health care facilities (such as hospitals and nursing homes) participating in the Medicare and Medicaid programs.b CMS also administers other programs designed to improve the quality of health care provided to Medicare and Medicaid beneficiaries as well as other patients.
Food and Drug Administration (FDA)	Part of FDA's responsibility for protecting the public health involves ensuring the safety and efficacy of human drugs and medical devices marketed in the United States, including antibiotics and diagnostic tests. FDA reviews and approves drugs, reviews and authorizes marketing for diagnostic devices, and provides educational information to consumers and health care providers about the appropriate use of antibiotics. FDA also approves, regulates, and collects data on antibiotics used in food animals, and monitors antibiotic-resistant bacteria in retail meat and poultry.
National Institutes of Health (NIH)	NIH's mission is to apply fundamental knowledge about living systems to enhance health, lengthen life, and reduce illness and disability. According to NIH officials, NIH and its National Institute of Allergy and Infectious Diseases (NIAID) fund and conduct research to better understand how antibiotic resistance develops, and supports the research and development of new diagnostic tests, treatments, and vaccines for antibiotic-resistant infections. For example, NIAID's Antibacterial Resistance Leadership Group provides extramural funding to a global consortium of scientific experts leading clinical research on important scientific questions related to antibiotic resistance.
Office of Global Affairs	The Office of Global Affairs represents the United States in international negotiations that set the worldwide agenda to address antibiotic resistance and coordinates across federal agencies to inform policies and programs. The office co-chairs the Transatlantic Task Force on Antimicrobial Resistance on behalf of the Department of Health and Human Services.

Source: Department of Health and Human Services documentation and prior GAO reports. | GAO-20-341.

[a]CARB-X restricts its funding to projects that target drug-resistant bacteria identified by CDC as a serious or urgent threat in its 2013 Threats Report, or by the World Health Organization as a critical or high threat.

[b]CMS refers to these health and safety requirements as Conditions of Participation or Conditions for Coverage, depending on the type of health care facility.

CDC Has Expanded Surveillance of Antibiotic Resistance, but Faces Challenges Determining the Magnitude of the Problem

Since the National Action Plan was released in 2015, CDC has made progress in expanding surveillance for antibiotic resistance in the United States and abroad. However, the magnitude of the problem and its trends over time remain unknown, in part because of challenges in three areas: (1) tracking antibiotic resistance across all health care settings, (2) reporting complete and timely information on magnitude and trends of antibiotic resistance, and (3) tracking and assessing the global antibiotic resistance threat.

CDC Has Expanded Surveillance of Priority Bacteria

To better assess the full extent of antibiotic resistance, CDC has expanded its surveillance of priority bacteria in the United States in order to better assess the full extent of antibiotic resistance since the 2015 National Action Plan was released.

CDC tracks antibiotic resistance through several infectious disease surveillance systems in collaboration with state and local health officials, health care providers and facilities, and laboratories.[28]

Rather than establishing a single surveillance system for antibiotic resistance, CDC generally incorporates tracking of antibiotic resistance into broader surveillance systems, according to agency officials. The surveillance systems are spread across various divisions within CDC that specialize in specific types of infection or certain settings. (See Table 2 for a description of each system and the resistant bacteria it tracks).

---

[28] Public health surveillance is the ongoing, systematic collection, analysis, and interpretation of health-related data essential to planning, implementation, and evaluation of public health practice.

Table 2. Description of Center s for Disease Control and Prevention (CDC) surveillance systems and priority bacteria tracked

Surveillance system	Description of system	Priority bacteria tracked
Antibiotic Resistance Laboratory Network (AR Lab Network)	The AR Lab Network provides nationwide lab capacity to rapidly detect antibiotic resistance in health care, food, and the community, and inform local responses to prevent spread and protect people. It includes labs in 50 states, several large cities, and Puerto Rico, including seven regional labs and the National Tuberculosis Molecular Surveillance Center.	Testing for: *Neisseria gonorrhoeae* *Acinetobacter* Carbapenem-resistant Enterobacteriaceae *Salmonella*, and more
Emerging Infections Program (EIP)	A network of public health-academic hospital collaborations in 10 states. It provides access to bacterial and fungal samples for testing and detailed clinical case data.[a] EIP has population level data, and does not focus on infections occurring in health care facilities as does the National Healthcare Safety Network. The three main programs within EIP collect different types of resistance data: • Active Bacterial Core provides clinical information and resistance data for bacteria that cause infections predominately in the community • The Healthcare-Associated Infections-Community Interface provides clinical information and resistance data for bacteria and fungi that cause infections at the intersection of health care and the general community Foodborne Diseases Active Surveillance Network (FoodNet) supplies clinical and epidemiologic data on the human enteric isolates followed in the National Antimicrobial Resistance Monitoring System (NARMS)	Active Bacterial Core: *Streptococcus pneumoniae* Groups A and B *Streptococcus* Healthcare-Associated Infections-Community Interface: *Clostridioides difficile* (*C. difficile*) Carbapenem-resistant Enterobacteriaceae Extended-spectrum Beta-lactamase- producing Enterobacteriaceae Carbapenem-resistant *Acinetobacter* Carbapenem-resistant *Pseudomonas aeruquinosa* (2016-2018) Methicillin-resistant *Staphylococcus aureus* (MRSA)[b] FoodNet: (see NARMS list)
Gonococcal Isolate Surveillance Program (GISP)	A program to track antibiotic resistance data for gonococcal isolates. Isolates are collected from sexually-transmitted disease clinics in approximately 28 cities.	*Neisseria gonorrhoeae*

Surveillance system	Description of system	Priority bacteria tracked
National Antimicrobial Resistance Monitoring System (NARMS)	A national public health surveillance system that tracks changes in the susceptibility of foodborne and other enteric bacteria to antibiotics of human and veterinary medical importance. NARMS is a collaborative effort among CDC, the Food and Drug Administration (FDA), the U.S. Department of Agriculture, and state and local health departments. CDC tests bacteria isolated from human samples, while FDA and the U.S. Department of Agriculture test bacteria isolated from samples taken from retail meats and food animals at slaughter, respectively.	*Salmonella Campylobacter Shigella Escherichia coli* O157
National Healthcare Safety Network (NHSN)	A system that collects and provides data on infections and drug resistance in health care settings. Since NHSN collects data directly from health care facilities, it can provide facility-level information on health care-associated infections and antibiotic resistance to those facilities and to CDC.	*Staphylococcus aureus Enterococcus* Enterobacteriaceae *Acinetobacter Pseudomonas aeruginosa C. difficile* MRSA
National Notifiable Diseases Surveillance System	CDC receives data on diseases that are deemed "nationally notifiable" by the Council of State and Territorial Epidemiologists. Notifiable disease surveillance begins at the level of local, state, and territorial public health departments (also known as jurisdictions). Jurisdictional laws and regulations mandate reporting of cases of specified infectious and non-infectious conditions to health departments.	Carbapenemase-producing Carbapenem- resistant Enterobacteriaceae Vancomycin-resistant *Staphylococcus aureus*
National Tuberculosis Surveillance System	Collects data on tuberculosis cases, including resistance data. Public health departments from 50 states and the U.S. territories contribute data.	*Mycobacterium tuberculosis*

Source: GAO summary of information from CDC websites. | GAO-20-341.

[a]CDC also tracks a fungal infection, fluconazole-resistant Candida, in the Emerging Infections Program.

[b]According to CDC, MRSA was formerly a part of the Active Bacterial Core.

According to CDC and other officials and documents we reviewed, including the National Action Plan Year 3 Progress Report, CDC has taken the following actions, among others, to expand surveillance in order to better assess the scope of antibiotic resistance:

- Established the Antibiotic Resistance Laboratory Network in 2016 to improve testing capacity to better identify antibiotic resistance in the United States. The network consists of 55 state and local (including Puerto Rico), and seven regional, public health laboratories and the National Tuberculosis Molecular Surveillance Center.[29] The network is improving and expanding laboratory capacity response at public health laboratories around the country, as well as at regional centers, according to representatives from two national professional organizations of state and local health officials and epidemiologists.
- Expanded antibiotic resistance-related efforts in its Emerging Infections Program (EIP), a network that seeks to monitor, prevent, and control emerging infectious diseases. For example, since 2015, more of the existing 10 EIP sites are conducting surveillance for invasive *Staphylococcus aureus* infections, carbapenem-resistant Enterobacteriaceae, and *C. difficile*, among others. Separately, the National Action Plan had included a goal for CDC to expand EIP by adding up to 10 sites within 3 years. However, CDC officials told us that in light of resource limitations, they chose to instead increase the number of pathogens reported at existing EIP sites. They told us they determined this was a better use of the limited funds, and that existing EIP sites are sufficient for current EIP efforts related to antibiotic resistance.

---

[29] The seven regional laboratories in the network are located in Maryland, Minnesota, New York, Tennessee, Utah, Washington, and Wisconsin. The laboratories were established under CDC's new efforts through the Antibiotic Resistance Solutions Initiative in 2016 to help health departments nationwide tackle antibiotic resistance and other patient safety threats, including through the Epidemiology and Laboratory Capacity for Infectious Diseases Cooperative Agreement.

- Updated the domestic tuberculosis surveillance system by incorporating advanced drug susceptibility testing and reporting and by developing capacity for state surveillance systems to report their tuberculosis test data electronically to CDC laboratories.
- Supported state and local health departments to better track, investigate, and prevent resistant foodborne disease, among other things, through the National Antimicrobial Resistance Monitoring System for Enteric Bacteria (NARMS). For example, the system can now carry out whole genome sequencing for all the pathogens it tracks, which enhances its detection and response capabilities, such as by expanding CDC's ability to detect new and emerging resistance, according to CDC officials.
- Launched the Enhanced Gonococcal Isolate Surveillance Program (eGISP), which augments the main Gonococcal Isolate Surveillance Program (GISP). Whereas GISP only collects samples from the urethras of men with symptoms of gonorrhea, in select sexually transmitted disease clinics, eGISP also collects samples from women and from other sites on the body, such as the throat. The specimens are sent to regional laboratories for resistance testing.[30]

CDC has also worked with international partners to expand surveillance of antibiotic resistance abroad. These efforts involved CDC collaborations with WHO, the European Center for Disease Prevention and Control, the government of the United Kingdom, other governments, and other multi- country efforts, such as the Surveillance and Epidemiology of Drug- Resistant Infections Consortium and the Transatlantic Taskforce on Antimicrobial Resistance (TATFAR).[31] The collaborations aimed to

---

[30] Regional laboratories for gonorrhea resistance testing as part of GISP and eGISP are located in Maryland, Tennessee, Texas, and Washington.

[31] The UK-based charitable organization the Wellcome Trust launched the Surveillance and Epidemiology of Drug-Resistant Infections Consortium to help countries track, share, and analyze information about antimicrobial resistance. The consortium brings together international experts in infectious diseases, epidemiology, and human and animal health to identify gaps in surveillance, help countries strengthen and sustain their capacity to collect data on drug-resistant pathogens, and improve global coordination. The group also provides

develop technical guidance to help improve surveillance in other nations and to organize an international forum. CDC also launched its Antibiotic Resistance (AR) Solutions Initiative, which invests in national and international infrastructure to address resistant infections across health care settings and communities and from food.

The Precise Magnitude and Trends of Antibiotic Resistance Are Unknown, in Part Because of Challenges CDC Faces in Three Areas

CDC faces three general challenges in tracking and reporting trends in antibiotic resistance. First, it faces limitations in data reporting and resistance testing from hospitals, as well as challenges ensuring that its resistant gonorrhea surveillance system is representative of the U.S. population. Second, CDC faces challenges in reporting complete and timely information on the magnitude of and trends in antibiotic resistance. Finally, CDC faces challenges to detecting resistance threats abroad.

Challenges in Tracking Resistance

The first challenge CDC faces in tracking trends in resistance is addressing low hospital participation in a new option of CDC's National Healthcare Safety Network (NHSN) system intended to address some limitations in NHSN. NHSN is, among other things, an online system for tracking health care-associated infections. It provides facilities, states, regions, and the nation with data needed to identify problem areas,

---

technical expertise and knowledge to help improve surveillance networks and look at how technology can be used to better understand resistance mechanisms and how infections spread. TATFAR was created in 2009 to address the urgent threat of antibiotic resistance. TATFAR's technical experts from Canada, the European Union, Norway, and the United States collaborate and share best practices to strengthen domestic and global efforts.

measure the progress of prevention efforts, and ultimately eliminate health care-associated infections, according to CDC.[32]

CDC established three modules within NHSN that allow hospitals to report select antibiotic-resistant infections, among other things, which include reporting required by states or by CMS, according to agency officials. Two modules track patients who have an infection associated with a medical device or resulting from a surgical procedure. Hospitals only report on resistance in these modules for specific combinations of antibiotics and bacteria, such as carbapenem-resistant Enterobacteriaceae. The third module tracks certain hospital patients who test positive for certain multidrug-resistant infections, including methicillin- resistant *Staphylococcus aureus* (MRSA)—a type of bacteria found on people's skin that is usually harmless but can cause serious infections, according to CDC.[33] However, according to CDC, many antibiotic- resistant infections detected during hospital care do not fall into one of these three modules and therefore would not be captured in NHSN, limiting CDC's ability to identify important new resistances or trends.

Health Care-Associated Infections

Patients in settings such as hospitals and long-term care facilities (e.g., nursing homes) in many cases already have a weakened immune system or an underlying illness, making an antibiotic-resistant infection especially dangerous, according to the Centers for Disease Control and Prevention (CDC). A high proportion of the morbidity and mortality associated with antibiotic resistance is seen in health care-associated infections. Tracking resistance in health care settings is therefore critical to national surveillance efforts.

Source: GAO summary of CDC documents. | GAO-20-341.

In 2014, to help address this limitation, CDC officials told us they introduced a new option for hospitals to report data on antibiotic resistance—the Antimicrobial Resistance Option (AR Option). This option

[32] CDC, VA, and DOD are collaborating to report some antibiotic resistance data to CDC via NHSN.

[33] The module tracks methicillin-resistant *Staphylococcus aureus* (MRSA), *C. difficile*, and carbapenem-resistant Enterobacteriaceae, among others.

allows for reporting of data on antibiotic resistance for certain bacteria, regardless of whether the patient has a health care-associated infection.[34] In contrast to the other three modules, reporting to the AR Option is voluntary.

As a result, while about 86 percent of the 17,529 eligible U.S. health care facilities participate in at least one of the older three antibiotic-resistance reporting modules, only about 10 percent of the 6,836 eligible hospitals participate in the newer, voluntary AR Option, according to our analysis of NHSN hospital participation data as of January 2020.[35] The hospital participation rate among U.S. states and territories ranged from no participation (in nine states and territories) to about 27 percent.[36] Representatives from a national association of state public health officials we interviewed said that this low rate limits the value of the data, a view that echoed the findings of a 2018 report by the Joint Public Health Informatics Task Force.[37]

CDC officials acknowledged that participation in the AR Option is low and cited reasons for this, including hospital resource limitations, and—in many cases because participation is voluntary—because hospitals do not prioritize submitting data to the AR Option. According to CDC officials, it is particularly challenging for many smaller hospitals and Indian Health Service facilities with resource constraints to participate, as it requires

---

[34] Hospital types in NHSN include general acute care hospitals, critical access hospitals, children's hospitals, oncology hospitals, long-term acute care hospitals, and inpatient rehabilitation facilities. The AR Option is part of a reporting module called the Antimicrobial Use and Resistance Module (AUR Module) that also tracks antimicrobial use through an "AU Option."

[35] Fewer facilities are eligible for the AR Option because that option does not extend to, for example, dialysis clinics or long-term care facilities. These totals do not include military bases outside the United States.

[36] Our analysis found there were no hospitals in the five U.S. territories participating in the AR Option as of June 2019, although some hospitals were enrolled in NHSN and eligible to participate in the AR Option.

[37] The Joint Public Health Informatics Task Force is a coalition of nine national public health associations, co-chaired and staffed by the Association of State and Territorial Health Officials and the National Association of County and City Health Officials, that helps U.S. governmental public health agencies build modern information systems across a spectrum of public health programs.

significant information technology investment.[38] The Joint Public Health Informatics Task Force report noted two other common challenges: low capacity for information technologies needed to support data submission to the AR Option, and a lack of motivated leadership, such as a facility "champion," to oversee the development and maintenance of needed reporting infrastructure. For example, the maintenance of reporting infrastructure could address changes to electronic medical records that are not immediately compatible with the AR Option reporting format.

CDC officials told us the agency is taking some steps to increase participation in the AR Option. For example, it is encouraging the over 1,500 hospitals (as of December 31, 2019) that are participating in a related reporting effort—known as the Antimicrobial Use Option (AU Option)—but not in the AR Option to participate in both. In addition, the agency is working with vendors of equipment and electronic health record software to make it easier for hospitals to participate in the AR Option.

One of CDC's goals for the AR Option is to use reported data to conduct regional and national assessments of resistance. To help meet this goal, officials said they would like participation by all eligible hospitals in the AR Option, but they have not determined the needed participation rates or appropriate distribution of participating hospitals. Our past work has shown that leading practices for federal strategic planning include articulating specific goals, establishing a method to assess progress toward these goals, and aligning the plans and goals with the agency's mission.[39]

---

[38] Beginning 2018, facility participation in the AUR Module was incentivized through making it an option for public health registry reporting under the Promoting Interoperability Program established by CMS to encourage clinicians and eligible hospitals, including critical access hospitals, to adopt, implement, upgrade, and demonstrate meaningful use of certified electronic health record technology. CDC officials told us that the AUR Module requires all data to be reported electronically, because manual reporting of these data would be time and staff prohibitive.

[39] We have previously reported that the strategic planning practices required at the federal agency level by the Government Performance and Results Act of 1993 (GPRA) and the GPRA Modernization Act of 2010 can serve as leading practices for planning at lower levels within agencies such as individual programs or initiatives. GAO, *Performance.gov: Long-Term Strategy Needed to Improve Website Usability*, GAO-16-693 (Washington, D.C.: Aug. 30, 2016); Government Performance and Results Act of 1993, Pub. L. No. 103- 62, 107 Stat. 285 (Aug. 3, 1993) and GPRA Modernization Act of 2010, Pub. L. No. 111-352, 124 Stat. 3866 (Jan. 4, 2011). These practices and associated Office of Management and Budget (OMB) guidance, together with practices we have identified, provide a framework

By taking steps to determine the participation rates and distribution of participation hospitals needed for CDC to meet its goal of conducting regional and national assessments of antibiotic resistance of public health importance, CDC would have more reasonable assurance that it can achieve its goal.

The second challenge CDC faces is ensuring representativeness of its resistant gonorrhea surveillance system. CDC has classified resistant gonorrhea as one of the most urgent antibiotic-resistance threats in the nation, affecting over half a million patients annually. According to the agency, resistant gonorrhea warrants this designation because of the limited remaining treatment options, the high number of gonorrhea infections, potential adverse outcomes (such as increased transmission of HIV), and the prospect that gonorrhea may become incurable if new resistance arises and spreads.

It is not clear, however, that GISP data are representative of the general U.S. population because GISP draws on a limited sample of that population.[40] Specifically, GISP collects culture specimens—-called isolates—and accompanying epidemiologic data from only the first 25 men with inflammation of the urethra consistent with gonorrhea visiting each participating sexually transmitted disease clinic each month. It does not collect culture specimens from women. In addition, the number of participating clinics each year has varied from 21 to 30 (see Figure 2 for the current sites).[41] CDC estimates that the cases of gonorrhea identified through GISP surveillance represent only about 1 to 2 percent of all reported cases of gonorrhea in the United States each year. Further, the

---

of leading practices in federal strategic planning. OMB defines a strategic goal as a "statement of aim or purpose" that "articulate[s] clear statements of what the agency wants to achieve." OMB Circular No. A-11, *Preparation, Submission, and Execution of the Budget*, Pt. 6, § 200.22 (Washington, D.C.: June 28, 2019).

[40] GISP was established in 1986 not only to monitor resistance trends in gonorrhea, but also to provide a basis for the selection of therapies to treat gonorrhea. GISP data are thus important for informing treatment recommendations, which can directly affect health care provider prescribing practices.

[41] According to CDC officials, each sentinel site has used a standard protocol, sampling methods, and culture-based laboratory methods consistently since GISP's inception, allowing for interpretation of trends over time. Sentinel surveillance is the study of disease rates in a specific cohort, such as in a geographic area or population subgroup to estimate trends in a larger population.

GISP sample design also over-represents cases in the western United States, where antibiotic-resistant gonorrhea has tended to initially emerge, according to CDC.[42] According to CDC, this design allows for more rapid detection of emerging resistance by ensuring a sufficient sample size from the western United States because resistance tends to emerge from that area. CDC has two projects—Strengthening the United States Response to Resistant Gonorrhea (SURRG) and eGISP— intended to, among other things, enhance domestic gonorrhea surveillance and learn more about the representativeness of GISP through limited testing of women and of body sites other than urethras, respectively.[43]

The Urgent Threat of Resistant Gonorrhea

According to the Centers for Disease Control and Prevention (CDC), gonorrhea is the second most commonly reported notifiable disease in the United States, with over 500,000 infections reported in 2017. However, CDC estimates that the true number could be as many as 820,000 each year. In addition to being a very common infection, gonorrhea is developing resistance to treatment options. As recently as 2006, CDC had five recommended options, but it estimates that nearly half of U.S. infections are now resistant to available antibiotics, including combinations. Consequently, it now recommends only one regimen. In 2014, a case of dual-therapy failure was reported in the United Kingdom, and in February 2018, a similar case in the United Kingdom was reported that also failed to respond to the last-resort therapy, spectinomycin, resulting in treatment failure. As of June 2019, CDC reported that it had not received any reports of verified clinical treatment failures to any cephalosporin in the United States.

Source: GAO summary of CDC and other information. | GAO-20-341.

---

[42] According to CDC, isolates from the western United States are over-represented in GISP compared with the geographic distribution of nationally reported gonococcal infections in men. For example, CDC reported that in 2014, isolates from the West represented 40 percent of GISP isolates but less than 25 percent of reported cases.

[43] SURRG is a project to enhance domestic gonorrhea surveillance and infrastructure and build capacity for rapid detection and response to resistant gonorrhea, among other things. eGISP was mentioned earlier in this report.

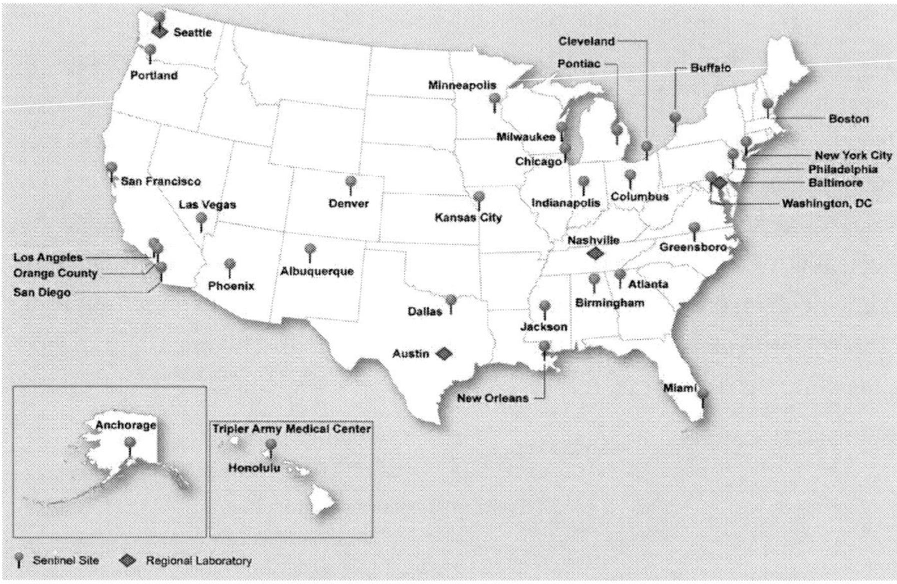

Source: Centers for Disease Control and Prevention. | GAO-20-341.

Figure 2. Resistant Gonorrhea Sentinel Surveillance Sites and Regional Testing Laboratories, 2018.

Data table for Figure 2: Resistant gonorrhea sentinel surveillance sites and regional testing laboratories, 2018

Sentinel Site
• Albuquerque
• Anchorage
• Atlanta
• Birmingham
• Boston
• Buffalo
• Chicago
• Cleveland
• Columbus
• Dallas
• Denver
• Greensboro
• Indianapolis
• Jackson
• Kansas City
• Las Vegas

- Los Angeles
- Miami
- Milwaukee
- Minneapolis
- New Orleans
- New York
- Orange County
- Philadelphia
- Phoenix
- Pontiac
- Portland
- San Diego
- San Francisco
- Seattle
- Tripler Army Medical Center, Honolulu
- Washington D.C.

Regional Laboratory

- Seattle
- Austin
- Nashville
- Baltimore

Note: Sentinel surveillance is the study of disease rates in a specific cohort, such as in a geographic area or population subgroup to estimate trends in a larger population. In the Gonococcal Isolate Surveillance Program, selected demographic and clinical data are abstracted from medical records, and isolates are tested for antibiotic susceptibility at regional laboratories.

However, CDC's current methodology may limit its ability to establish a representative trend. According to CDC officials, GISP could improve its representativeness by adding clinics or covering more of the population at its current sites. However, efforts to expand GISP would be difficult due to limited local capacity (see text box). CDC has taken some steps to assess the representativeness of the current GISP design, but it has not conducted a comprehensive study to assess the representativeness of the trends identified in GISP. A 2015 CDC evaluation concluded that the representativeness of GISP was "good" on a scale of fair, good, or great.[44]

---

[44] A 2007 evaluation of GISP noted that only males are represented; private practice patients are not included; the system relies on urethral cultures, so isolates from pharyngeal and rectal cultures are not represented; only 25-30 cities are represented; and these cities are weighted toward the West Coast.

Barriers to Expanding the Gonococcal Isolate Surveillance Program (GISP)

GISP currently tracks a limited sample of the U.S. population. According to Centers for Disease Control and Prevention (CDC) officials, a more thorough expansion of GISP would be more difficult because of limited local capacity to conduct culture-based testing for resistance in gonorrhea. Specifically, laboratories increasingly use newer gonorrhea testing technology that gives more rapid results but cannot currently be used to test for resistance. This trend has contributed to the reduced capability of many laboratories to perform the gonorrhea culture-based testing for antibiotic susceptibility testing, to the point that many clinics cannot collect specimens for testing, according to CDC officials. Furthermore, officials said that adding new clinics to GISP would require financial and other resources for, among other things, establishing culture testing for resistance and information technology needed to report data to the system.

Most gonorrhea cases are diagnosed outside sexually transmitted disease clinics. However, expanding GISP to non-sexually transmitted disease clinic sites could be particularly costly and inefficient, officials said, because these sites tend to see many fewer gonorrhea cases per year compared to sexually transmitted disease clinics; therefore they may not be able to contribute significant data to GISP. Through the Strengthening the United States Response to Resistant Gonorrhea (SURRG) project, CDC is currently exploring options to work with states to enhance gonorrhea testing capacity. This program was established in 2016 but has not received the funding needed to expand capacity to the extent CDC had planned. In addition, physicians and other providers have limited time to devote to data collection and reporting needed to participate in GISP. CDC officials also told us the reimbursement rates for providers for these services are inadequate.

Source: GAO summary of CDC documents and interview with CDC officials. | GAO-20-341.

However, the evaluation covered only part of fiscal year 2014 and consisted of a limited comparison of selected demographic characteristics captured in gonorrhea cases identified in GISP to those captured through the National Notifiable Diseases Surveillance System, according to CDC officials, and which has its own limitations. Further, the results of this evaluation have not resulted in any changes to the GISP design. CDC officials told us they hope to learn more about the representativeness of GISP urethral isolates from testing women, patients in non-sexually transmitted disease clinic sites in the SURRG project and eGISP, and testing at other body sites, and then comparing some of these results to those of GISP. However, these efforts overall were not specifically designed to fully assess the representativeness of GISP and may not provide a sufficient assessment for impacting changes to the GISP design.

CDC's guidelines of efficient and effective public health surveillance systems state that, in order to be representative, the data from a public health surveillance system should accurately reflect the characteristics of the health-related outcome—such as resistant gonorrhea—under surveillance.[45] A more precise evaluation of the representativeness of the surveillance system can be done via carefully designed studies to obtain complete and accurate data for the health event in question—namely, the urgent threat of antibiotic-resistant gonorrhea. By evaluating the surveillance system for resistant gonorrhea to ensure that it includes measures of its representativeness, such as by comparing the trends in the sample population with those in the overall U.S. population, using specially designed studies if needed, CDC would have better assurance that the trends detected in GISP accurately reflect the characteristics of the health-related outcome the system is designed to monitor.

In addition to the limited design of GISP, CDC faces the challenge of competing priorities under reduced funding that precluded it from completing its plans to expand the SURRG project. The SURRG expansion was designed to address a National Action Plan goal of controlling resistant gonorrhea, among other things, but also affects surveillance, as CDC officials told us SURRG was established to address some limitations in GISP surveillance. Specifically, one of the plan's milestones assigned to CDC is to maintain advanced capacity for rapid response to antibiotic-resistant gonorrhea for at least 20 state health departments. Such capacity includes detection, diagnosis, and investigation of suspected resistant cases within their state or region and assistance for health care providers in appropriately treating infected patients. CDC officials told us that because they received about half of the appropriations they had requested, CDC had to make cuts in some of their projects, and

---

[45] These characteristics generally relate to time, place, and person. Department of Health and Human Services, Centers for Disease Control and Prevention, "Updated Guidelines for Evaluating Public Health Surveillance Systems," *Morbidity and Mortality Weekly Report*, vol. 50, no. RR-13 (2001).

SURRG was one of those that CDC chose to reduce.[46] Eight SURRG sites, rather than the 20 recommended by the National Action Plan, collect and analyze data. However, in its progress reports covering the first 4 years of the National Action Plan's implementation, the CARB Task Force did not identify plans to address barriers related to expanding the SURRG project.[47] The CARB Task Force coordinators told us that the progress reports have not identified plans to address barriers largely because the task force focused on reporting the agencies' accomplishments in implementing the National Action Plan. The coordinators also said that, in response to our inquiries during this review, the task force intends to identify agencies' plans for addressing barriers in the progress report to be published in fall 2020.

The Executive Order directs the CARB Task Force to provide annual updates to the President on federal government actions to combat antibiotic resistance, including progress made in implementing the National Action Plan, plans for addressing any barriers preventing its full implementation, and recommendations for any new or modified actions, taking federal government resources into consideration. Without reporting its plans to address such barriers, the CARB Task Force has not provided all the information required by the Executive Order and has not fully carried out its role to facilitate and monitor implementation of the National Action Plan, which may reduce the effectiveness of federal efforts to combat antibiotic resistance.

The third challenge CDC faces tracking antibiotic resistance is addressing limitations to the use of test results in surveillance in health care settings. For example, some health care facilities are not using the most up-to-date testing methods for determining whether the bacteria causing an infection are resistant to certain antibiotics, according to CDC officials and

---

[46] For example, CDC officials provided documentation establishing that it requested $264.3 million for fiscal year 2016 for combating antibiotic-resistant bacteria initiatives. CDC also provided documentation supporting that its fiscal year 2016 operating plan included about $160 million for these initiatives; thus, CDC chose not to allocate the resources to implement all activities outlined in the National Action Plan, including the full implementation of the SURRG sites.

[47] CARB Task Force, *National Action Plan: Progress Report for Years 1 and 2*, *National Action Plan: Progress Report: Year 3*, and *National Action Plan: Progress Report: Year 4*.

a report from the Antibiotic Resistance Surveillance Task Force.[48] In addition, laboratories may only report an interpretation of the test result to CDC (e.g., whether the bacteria is resistant or susceptible to an antibiotic) and not the quantitative results (e.g., measures of the growth of bacteria in the presence of the antibiotic). This presents a challenge for comparing data from different laboratories, since they may not be using consistent testing thresholds for determining antibiotic resistance. Another limitation is that some test equipment may be designed to give limited results for the purposes of guiding treatment recommendations and stewardship efforts, which may also limit the information available to CDC.[49] For example, the test may inform the user that the infection is susceptible to one antibiotic but "suppress" information on susceptibility to other antibiotics, in order to guide the user toward treatment with the preferred first-line treatment. The Antibiotic Resistance Surveillance Task Force report noted that some suppression is done by the testing equipment itself and some by software systems that record, manage, and store data for clinical laboratories. CDC officials told us they are working with some diagnostic test manufacturers to explore these issues and develop solutions to address them. The Antibiotic Resistance Surveillance Task Force is also working to address the diagnostic test challenges related to antibiotic resistance surveillance.

Challenges in Reporting Complete and Timely Information on Magnitude and Trends

CDC also faces challenges in reporting timely and complete information on the magnitude of and trends in antibiotic resistance in the agency's Threats Reports. One challenge is in providing information in these reports on the uncertainties in reported numbers of deaths from antibiotic- resistant infections. Another challenge is in issuing such reports in regular, timely intervals. As a result of these challenges, among others,

---

[48] The 2018 report from the Antibiotic Resistance Surveillance Task Force, led by the Council of State and Territorial Epidemiologists, indicated that although FDA has adopted new breakpoints for antibiotic resistance tests and equipment manufacturers are able to implement them, many clinical laboratories use some old breakpoints. The task force further noted that standardized and widely used up-to-date breakpoints are fundamental to quality of data and the effective and coordinated action based on the data.

[49] We discuss antibiotic stewardship later in this report.

the true magnitude of, and trends in, antibiotic resistance over time are unknown, including trends in various places and among people with various characteristics.

Surveillance for antibiotic resistance is complex and costly, according to experts at our meeting, CDC officials, and literature we reviewed. Experts told us such surveillance encompasses diverse pathogens, diseases, and health care settings and requires a variety of data sources and collection efforts. Furthermore, experts from our meeting told us the fundamental data required—such as data on the number of illnesses and deaths attributable to resistance and data on related health care costs—are currently insufficient. One expert added that there is a lack of real-time monitoring data, such as data that are available within hours or days of being generated. The data gaps are especially large for infections acquired in the community, as opposed to in a health care setting, because there is very limited tracking of such infections and whether they are resistant. As a result, CDC officials said, it is challenging to provide ranges of uncertainty, a critical component of any effort to measure and report on magnitude and trends.[50]

Neither the 2013 Threats Report nor the 2019 Threats Report provided quantitative measures of uncertainty, such as confidence intervals, for CDC's estimates of morbidity and mortality resulting from antibiotic-resistant infections. For example, the report stated that there are at least 23,000 deaths a year as a direct result of antibiotic-resistant infections, but it did not include an upper limit or a single point estimate for this number. Similarly, the 2019 Threats Report stated that there are at least 35,900 deaths a year, without an upper limit or a single point estimate. A recent re-estimate by a group of scientists has put the likely minimum number of

---

[50] One type of uncertainty range is the confidence interval, which gives a range of values that is likely to include the true value. In the case of CDC's 2013 Threats Report and the *Antibiotic Resistance Threats in the United States, 2019* report, confidence intervals would provide a range for the number of infections and deaths, along with a degree of confidence—usually expressed as a percentage—that the true values fall with thatrange.

deaths annually in the United States at approximately 153,000, or about four times the 2019 CDC minimum estimate.[51]

CDC officials told us that because of several limitations, its estimates were the best that could be derived from the data available. For example, for the 2013 Threats Report, CDC only had data from a national hospital survey intended to produce estimates of all health care-associated infections and indirect estimates of the proportion of infections that were resistant. These data did allow CDC to calculate confidence intervals for infections by specific pathogens, but this information was not disclosed in the Threats Reports. Because the data sources were not intended for this purpose, the 2013 intervals were wide, from approximately 26 percent to 380 percent of the point estimates for each pathogen. CDC officials told us they elected not to include these ranges of uncertainties to avoid confusion in the 2013 Threats Report, because the report was intended for a variety of audiences, including the general public. Officials told us they planned to provide confidence intervals in an appendix of the 2019 Threats Report, but they did not.[52] CDC officials explained that they elected not to include confidence intervals in the 2019 Threats Report because several publications are pending that provide more granular data for many of the estimates included in the report.[53] It is thus unclear whether CDC plans to include any measures of uncertainties in future Threats Reports.

Federal standards for agency dissemination of information it produces stipulate that when information products are disseminated, error estimates are calculated and disseminated to support assessment of the appropriateness of the uses of the estimates or projections.[54] Providing

[51] J. P. Burnham, M. A. Olsen, and M. H. Kollef, "Re-estimating Annual Deaths Due to Multidrug-Resistant Organism Infections," *Infection Control & Hospital Epidemiology*, vol. 40, no. 1 (2019): pp. 112-113. This recent estimate by a group of scientists has put the likely minimum number of deaths annually in the United States at approximately 153,000.

[52] Both the 2013 and 2019 CDC Threats Reports have an appendix that discusses sources of data and methodologies but does not provide estimates of uncertainties. In some cases, CDC disclosed ranges, such as for the attributable costs of certain *C. difficile* infections.

[53] CDC officials told us they wanted to publish CDC's methodology in a peer-reviewed journal to allow for stakeholder input. They said multiple publications, including articles containing confidence intervals for some pathogens in the 2019 Threats Report, are pending.

[54] Office of Management and Budget, *Standards and Guidelines for Statistical Surveys* (September 2006).

measures of uncertainties in antibiotic resistance estimates, such as standard errors or confidence intervals, as appropriate, in its Threats Reports would help CDC and others compare information within and across reporting efforts, without having to consult multiple documents over time. CDC and others could use this information to draw appropriate conclusions about the characteristics of antibiotic resistance in the United States, including limitations associated with reported findings and conclusions.

Additionally, CDC does not have a plan for timely, regular issuance of their Threats Reports. It took CDC over 6 years to update the 2013 Threats report. CDC officials told us this length of time between reports was in part because, following issuance of the 2013 Threats Report, the agency was focused on implementing priority actions to improve antibiotic resistance surveillance data, including those efforts prescribed by the National Action Plan. In some cases, implementing these actions involved new data collection efforts that took time to establish, including that it can take up to 2 years to get new surveillance variables cleared by the Office of Management and Budget (OMB), CDC officials told us. In addition, CDC officials said it is time consuming to coordinate across the decentralized structure of antibiotic-resistance tracking at CDC to compile a consolidated report.

However, lack of timely, regular updates may affect the information available to the public as well as policy-makers. For example, the 2013 Threats Report stated that there are at least 23,000 deaths a year as a direct result of antibiotic-resistant infections. The 2019 report stated the number of deaths each year to be at least 35,900 deaths a year. This report also revised the 2013 estimate from 23,000 to 44,000 deaths a year, suggesting a nearly two-fold revision to the initial 2013 estimate.

CDC officials told us they would like to publish the report more frequently than every 6 years, and that it is reasonable they would develop such a plan for frequency of publication following the 2019 report. However, they said the agency does not currently have a plan for how often it will release future consolidated reports.[55] CDC's attributes of efficient and effective public health surveillance systems include timely data dissemination for planning, implementing, and evaluating public health policies and programs.[56] By developing a plan for more frequent dissemination of consolidated reporting on priority pathogens at regular intervals, CDC would have more timely trend data and other information necessary for users of the data, including policymakers, to prioritize, plan, implement, and evaluate public health actions to address antibiotic resistance.

Challenges in Tracking and Assessing the Global Threat

Through interviews with WHO and CDC officials, experts at our meeting, and document reviews, we identified three key challenges to CDC's ability to detect antibiotic resistance threats from abroad.[57] These challenges are as follows:

- *Data completeness, quality, and representativeness.* Data on antibiotic resistance from the national surveillance systems of some countries are incomplete because of a lack of capability and resources for implementing standardized protocols, according to WHO officials. Moreover, most information on antibiotic-resistant infections is limited to laboratory test data and does not include epidemiological data, such as data on the patient and location,

---

[55] According to the 2013 Threats Report, CDC stated it would update the assessment of priority bacteria at least every 5 years, although it did not specify that the reporting of the priority bacteria would occur every 5 years.
CDC officials told us CDC shares antibiotic resistance data at regular intervals, such as through GISP reports, but these are for specific pathogens or subsets. This does not address the timing of regular reporting of CDC's consolidated reports, such as the 2013 and 2019 Threats Reports.

[56] Department of Health and Human Services, Centers for Disease Control and Prevention, "Updated Guidelines for Evaluating Public Health Surveillance Systems."

[57] International surveillance, including surveillance of antimicrobial resistance, is a key component of the 2018 U.S. National Biodefence Strategy. Specifically, the strategy states that "Goal 1: Enable risk awareness to inform decision-making across the biodefense enterprise" is to build risk awareness by ensuring that domestic and international biosurveillance and information-sharing systems are coordinated and are capable of timely bioincident prevention, detection, assessment, response, and recovery. The White House. *National Biodefense Strategy, 2018* (Washington, D.C: September 2018).

which could provide additional insight about the circumstances around the resistant infection. Also, a lack of a sampling strategy for the detection of cases that are antibiotic-resistant may bias the representativeness of the data and interpretation of results. Specifically, when case identification is done only on the population of patients that seeks medical care and is tested, or when testing of the population varies such as across health care settings, the incidence and trends determined from this population may not represent the total population of concern.

- *Aggregated data reporting.* Some countries report aggregated, rather than isolate, or infection-level, data to the WHO's Global Antimicrobial Resistance Surveillance System (GLASS), a practice that WHO officials stated creates a challenge for data analysis and results interpretation.[58] According to officials, such aggregation limits statistical analysis that can be performed and limits analysis of factors such as the specific antibiotic-resistant bacteria, or the age or gender of the patient, among other things.
- *Surveillance is a complex function.* Many different health care and public health professionals are involved in the multistep process for generating data, according to a WHO report on GLASS. According to WHO officials, obtaining the staff commitment and training needed to ensure high-quality data can pose a challenge to public health agencies and health care organizations.

As we noted above, CDC has worked with, and continues to work with, international partners to expand surveillance of antibiotic resistance abroad, including through U.S. participation in GLASS. For example, CDC has helped develop technical guidance for surveillance programs in other countries and has organized international forums for surveillance. CDC officials also told us portions of domestic surveillance systems data collection include collection of patient travel history.

---

[58] An expert from our meeting later told us that WHO has begun to collect some isolate- level data within the GLASS platform, and it is working to expand these isolate-level reporting modules.

The Global Antimicrobial Resistance Surveillance System

In October 2015, the World Health Organization (WHO) launched the Global Antimicrobial Resistance Surveillance System (GLASS). The objectives of GLASS are to foster national surveillance systems and harmonized global standards and estimate the extent and burden of antimicrobial resistance globally by selected indicators, among other things. As of November 2019, 86 countries were enrolled in GLASS, a 25 percent increase over 2018. Participants were in various stages of economic development (13 lower-income countries, 23 lower-middle- income countries, 17 upper-middle-income countries, and 33 high-income countries) and in all WHO regions. Seventy-five countries provided descriptive information on their surveillance systems for tracking antimicrobial resistance, and 57 countries provided resistance data for 2018.

Source: GAO summary of WHO information. | GAO-20-341.

Federal Agencies Have Helped Advance Diagnostic Tests and Promoted Their Use, but These Efforts Have Limitations

Federal agencies have helped advance the development of new FDA-authorized tests and the use of existing tests for diagnosing antibiotic-resistant infections, but these efforts have limitations. Specifically, HHS and DOD have funded studies and taken other steps to advance testing, but they have not defined leadership, roles, and responsibilities to address a key barrier to the use of tests: a lack of clinical outcome studies. FDA has taken additional steps to advance testing; however, it has not regularly monitored test updates.

Agency Efforts toward the Development and Use of Diagnostic Tests

HHS and DOD Have Funded the Development of New Tests

HHS and DOD have awarded grants and contracts for the development of new FDA-authorized tests for diagnosing antibiotic-resistant infections.[59] Some of these awards address specific needs in the current availability of FDA-authorized tests, while others support more general research and development efforts. In addition, these agencies have taken steps to help reduce the chances of duplicative funding. According to experts, tests for antibiotic resistance not only help clinicians decide what antibiotics to use, they also provide important information for surveillance, including the number of cases of resistant infections in a population and the mechanisms of resistance.

Challenges in Addressing Diagnostic Test Gaps

According to federal documents and literature, challenges to diagnostic test development include:

- Lengthy and costly regulatory requirements, including additional regulatory hurdles in other countries
- Limitations in technical feasibility assessments due to intellectual property protection and conflict of interest requirements
- Differences in expertise between manufacturers and regulatory bodies
- Limited evidence on cost-effectiveness and clinical outcomes for using tests
- Limited resources in some settings to transport specimens, conduct, and maintain tests

Source: GAO summary of Presidential Advisory Council on Combating Antibiotic-Resistant Bacteria and other documents. | GAO-20-341.

Agencies have funded the development of some tests to address critical needs. We identified three such needs through interviews with federal agency officials and with experts. Specifically, according to agency

[59] This report focuses on FDA-authorized tests, which account for the majority of tests used in the United States, according to experts. According to FDA officials, there are non- FDA-authorized diagnostic tests for resistance available for all bacteria, including all of those on the 2013 threats list. Non-FDA-authorized tests can only be conducted by laboratory personnel meeting proficiency standards set by the Clinical Laboratory Improvement Amendment. However, experts told us the majority of tests used in the United States are FDA-authorized tests. For example, one expert estimated that more than 95 percent of laboratories using culture-based tests—tests that examine how well bacteria grow in the presence of an antibiotic—in the United States use some FDA- authorized tests.

officials and experts, there are no FDA-authorized tests that can do the following:

- Detect resistant gonorrhea or resistant campylobacter infection, which CDC identified as urgent and serious threats, respectively, in both the 2013 and 2019 Threats Reports.
- Rapidly detect resistance for seven other priority bacteria, according to the 2013 Threats Report.[60]
- Differentiate between viral and bacterial infections. Such a test would be useful primarily in preventing use of antibiotics for viral infections, which can contribute to the development of resistance in bacteria, among other things.

HHS and DOD have awarded funding to address these needs. For example:

- CARB-X—a program supported by NIH and BARDA within HHS—has awarded funding to a company to develop a rapid test to both diagnose gonorrhea and test for antibiotic resistance.
- CARB-X is funding other companies to, among other things, develop rapid testing for identification of and resistance in bloodstream infections, including for some priority bacteria.
- In September 2016, NIH and BARDA announced the Antimicrobial Resistance Rapid, Point-of-Need Diagnostic Test Challenge. As of December 2019, there were five finalists, working on such projects as developing a rapid test to differentiate viral from bacterial infections and developing a test that can

---

[60] The seven priority bacteria are non-typhoidal *Salmonella*, *Salmonella* typhi, *Shigella*, *Streptococcus pneumoniae*, vancomycin-resistant *Staphylococcus aureus*, erythromycin-resistant Group A *Streptococcus*, and clindamycin-resistant Group B *Streptococcus.* FDA officials noted there is no test for resistant *C. difficile*, but we do not count this as a gap because CDC's priority bacteria list does not include resistant *C. difficile,* only *C. difficile.* There is no single federal definition of the term "rapid." It can range from minutes to hours, depending on context.

identify or detect antibiotic-resistant bacteria, including antibiotic-resistant gonorrhea.[61]

- Within DOD, the Defense Advanced Research Projects Agency officials told us that the agency used fiscal year 2015 funding on contracts for the development of rapid molecular tests for resistant gonorrhea and to distinguish between viral and bacterial infections.

Federal agencies have also funded more general research and development efforts related to resistance testing. For example:

- NIH officials told us their agency has supported extramural projects related to the development of tests for antibiotic resistance by issuing grants and entering into contracts since fiscal year 2015.
- Separately from the Antimicrobial Resistance Diagnostic Challenge, BARDA entered into contracts with three organizations to develop tests focusing on the advanced stages of test development, including clinical trials, according to BARDA officials.
- Within DOD, the Defense Threat Reduction Agency is funding three projects using Other Transaction Authority or direct funding to a DOD Service laboratory, for developing tests.[62]

Federal agencies have also taken steps to help reduce the chances of duplicative funding, including working with some international efforts to develop tests, according to agency officials. For example, NIH reviews current and pending support of key project personnel prior to issuing of any research award, to help ensure NIH support complements support from other agencies and organizations. Similarly, officials from HHS's Office of Global Affairs worked during the creation and launch of the NIH- BARDA

[61] According to NIH officials, NIH is responsible for the overall coordination, oversight, and management of the challenge. It is anticipated that all tests will have a maximum time to result of 90 minutes.

[62] Other Transaction Authority, in this case, is the term used by DOD to refer to legal acquisition instruments other than contracts, grants, or cooperative agreements, to carry out certain prototype projects. 10 U.S.C. § 2371b.

challenge and an analogous United Kingdom innovation foundation competition called the Longitude Prize to help ensure these programs were designed to support different aspects of needed diagnostics.[63]

HHS Has Funded Some Studies of Clinical Outcomes, but Has Not Clearly Identified Leadership, Roles, and Responsibilities

HHS has funded some studies to assess the extent to which testing patients to identify whether they have antibiotic-resistant infections leads to improved clinical outcomes, such as more effective treatment for patients or more judicious use of antibiotics.[64] However, HHS has not identified relevant leadership, roles, and responsibilities among the HHS agencies that could fund such studies.

Clinical outcome studies are important for encouraging the use of diagnostic tests for antibiotic resistance, among other things, because such studies can demonstrate the benefits of those tests. According to PACCARB, there is very limited information on why clinicians sometimes forgo diagnostic testing, but one possible explanation is that there may be limited data demonstrating the value of such testing. In the absence of such data, a clinician may choose to treat the patient immediately rather than using a test for antibiotic resistance that has unknown value. Research into the clinical outcomes associated with such testing could therefore be used to help promote the use of those resistance tests that are found to be beneficial. As a result, patient care could be improved and clinicians could be guided towards appropriate antibiotics to prescribe.

Two HHS agencies have awarded grants for studies on the clinical outcomes of resistance testing, according to agency officials. For example, NIH provided grant support for a study that found, among other things, that using a rapid blood test for a range of potential bacteria and antibiotic

---

[63] The Longitude Prize, run by a United Kingdom-based innovation foundation, is a 10 million pound prize with an 8 million pound payout intended to foster the development of an affordable, rapid, point-of-care test that will conserve antibiotics and improve delivery of global health care. See "Longitude Prize," accessed May 22, 2019, https://longitudeprize.org.

[64] Previously, we reported that clinical benefits have not been well-established for some types of tests for infectious diseases. GAO, *Medical Devices: Challenges and Capabilities to Enable Rapid Diagnoses of Infectious Diseases*, GAO-17-347 (Washington, D.C.: Aug. 14, 2017).

resistance led to more judicious use of antibiotics.[65] Similarly, officials from the Agency for Healthcare Research and Quality (AHRQ) stated that the agency is funding investigator-initiated grant studies to assess the impact of tests on antibiotic stewardship. However, agency officials only mentioned these and a few other examples of studies they have funded on clinical outcomes.

International Needs for Diagnostic Tests for Antibiotic Resistance

To better understand international needs for antibiotic resistance tests, we interviewed officials from international organizations and the Office of Global Affairs within the Department of Health and Human Services (HHS). A Public Health England official told us that United Kingdom users are not confident that these tests will have a clinical impact or be cost effective. Similarly, an official from a trade organization of British medical test manufacturers told us that the value of tests for antibiotic resistance needs to be captured and disclosed, especially because people are more willing to pay for treatment than for tests.

However, other factors could also be important in determining which tests will be useful internationally. World Health Organization officials told us that they are working to determine what characteristics health care providers worldwide identify as key to making tests useful, so industry can develop such tests. They noted that tests designed for use in the United States may not be suitable for use in other countries. They also noted that laboratories in developing countries may not have the capacity to culture bacteria, so many need to use culture- independent tests.

Office of Global Affairs officials told us that a big challenge is developing accessible tests for use internationally. Their ideal test would be inexpensive, rapid, and capable of point-of-care use. They noted that cost and usability are the barriers to test use, not technology, and that use of existing tests remains limited, including within the United States.

Source: GAO summary of interviews with HHS and international organizations | GAO-20-341.

[65] R. Banerjee et al., "Randomized Trial of Rapid Multiplex Polymerase Chain Reaction- Based Blood Culture Identification and Susceptibility Testing," *Clinical Infectious Diseases*, vol. 61, no. 7 (2015): pp. 1071-1080, and E. L. Tsalik et al., "Advancing Diagnostics to Address Antibacterial Resistance: The Diagnostics and Devices Committee of the Antibacterial Resistance Leadership Group," *Clinical Infectious Diseases*, vol. 64, no. S1 (2017): pp. S41-S47.

Agency officials and experts agree that more needs to be done to evaluate clinical outcomes associated with use of diagnostic tests for antibiotic resistance. For example, in 2017, PACCARB reported that "there is a lack of clinical and economic outcome studies showing that any diagnostic test could prevent the emergence of antibiotic-resistant bacteria and would be cost effective."[66] Officials we interviewed from AHRQ, BARDA, CDC, FDA, and NIH all agreed with that PACCARB statement. Additionally, experts told us that such studies are lacking but important for advancing the use of tests. For example, one health care organization official told us the decision to adopt a test is based at least in part on whether there will be a clinical benefit. An infectious disease expert noted that to provide incentives for test use there needs to be some evidence that tests affect and improve care, but that most tests do not come with any evaluation of how they perform in practice. International organizations expressed similar opinions.

One reason for the relatively low number of studies is that those agencies that could conduct or fund diagnostic outcome studies have not clearly identified leadership, roles, and responsibilities for doing so. Although they agree that more such studies are needed, they have not identified which agency or agencies should take the lead, and what the roles of the other agencies should be. Instead, agencies have offered differing views on what each agency could do. For example, BARDA officials told us their agency has not funded such studies because it generally does not play a role in test adoption. BARDA officials, as well as officials from DOD and NIH, said that CDC should play a role in funding or conducting the studies. However, CDC officials told us that a lack of resources has prevented their agency from doing so, and that the responsibility should fall at least partly on BARDA.

Our previous work shows that key practices for interagency collaboration include identifying a lead agency (or, if leadership is shared, clearly identifying roles and responsibilities among the lead agencies), as

---

[66] Presidential Advisory Council on Combating Antibiotic-Resistant Bacteria, *Recommendations for Incentivizing the Development of Vaccines, Diagnostics, and Therapeutics to Combat Antibiotic Resistance* (Washington, D.C.: September2017).

well as clarifying the roles and responsibilities of all participating agencies.[67] By taking these actions, agencies—including AHRQ, BARDA, CDC, FDA, and NIH—could more effectively address the need for clinical outcome studies. Those studies, in turn, could help demonstrate the value of diagnostic tests for antibiotic resistance, potentially increasing their use, improving patient care, and enhancing stewardship efforts.

CMS and FDA Have Taken Steps to Advance the Use of Tests, but Experts Have Identified Challenges with Payments

CMS and FDA have taken some steps to advance the use of tests, including those to identify antibiotic-resistant bacteria. For example, FDA established a Payor Communication Task Force, which helps facilitate communication between test manufacturers and payors.[68] Such communication is important because payors decide whether tests will be covered by insurance, among other things. According to an FDA web page, by communicating with payors, test manufacturers could, for example, learn what data payors need to approve a test for coverage and then use this information to design clinical trials to provide that information. This process could reduce the time between when a test is cleared or approved by FDA and when it is covered.[69]

A similar step FDA and CMS took to advance the use of tests was to extend the Parallel Review program indefinitely, a move they announced in 2016. This program established a mechanism for FDA and CMS to simultaneously review clinical data, with the aim of reducing the time

---

[67] GAO, *Managing for Results: Key Considerations for Implementing Interagency Collaborative Mechanisms,* GAO-12-1022 (Washington, D.C.: September 27,2012).

[68] According to an FDA web page, payors include public payors such as CMS, private health plans, health technology assessment groups, and others who provide input into coverage, procurement, and reimbursement decisions. See Food and Drug Administration, "Payor Communication Task Force," accessed September 26, 2019, https://www.fda.gov/about-fda/cdrh-innovation/payor-communication-task-force. FDA told us this task force was established approximately 5 years ago.

[69] CMS officials told us they prefer the term "payment" to describe what documents such as the National Action Plan and PACCARB report call "reimbursement."

between FDA's approval and CMS's decision on whether to pay for the test.[70]

Experts told us challenges remain with test payments that may result in lower test use. For example, a PACCARB report states that "currently, [payment] for many diagnostic tests is not aligned with the value of the test," and noted that supplementing payments for tests could drive test development and use. BARDA officials also told us that a major factor affecting adoption of new tests is the cost of the test relative to reimbursement. Additionally, experts, including those at our meeting, told us that test payments remain insufficient to encourage broad test use.[71] For example, two experts from our meeting said that there is not always a clear link between the medical value of a test and the payment level for that test. One of these experts added that their laboratory decided not to adopt a test because low payment levels relative to costs made doing so a money-losing proposition. Three other experts we interviewed agreed that disparities between cost and payment can discourage test adoption.

Regarding federal payments for tests involving CMS and their payments through Medicaid and Medicare, there are limits to CMS's ability to address any disparities. For example, CMS officials told us the payments for some tests are based on a weighted, median, private-payor rates pursuant to the Protecting Access to Medicare Act of 2014, so CMS cannot specify the methodology used to set those rates.[72] Further, for inpatient tests, Medicare pays hospitals a single, bundled payment per patient stay, which is based on multiple factors, including the patient's diagnosis and treatment strategy, rather than on a specific service. As such, a separate payment for individual tests is not made under Medicare.

---

[70] This mechanism would pertain to tests receiving marketing authorization as de novo devices, but not to tests cleared under the 510(k) process based on a determination of substantial equivalence to an already legally marketed test.

[71] We did not independently verify expert claims regarding payments for test use.

[72] Pub. L. No. 113-93, § 216, 128 Stat. 1040, 1053 (codified in pertinent part at 42 U.S.C. § 1395m-1(b)).

FDA Efforts to Advance the Development of New Tests

FDA Has Taken Steps to Speed the Development of Tests for Newly Approved Antibiotics

FDA has taken steps toward the development of FDA-authorized tests for resistance for newly approved antibiotics—a process that currently can take months to years, according to experts and agency officials.[73] The delay stems in part from the need for a critical testing threshold known as a breakpoint—the threshold that is used to help a clinician decide whether or not a pathogen is resistant to the antibiotic (see text box).[74] The breakpoint of a new antibiotic is generally finalized only when FDA has approved the antibiotic. This means that breakpoints may often not be available for test manufacturers until after a new antibiotic is FDA- approved.[75] As a result, test manufacturers generally may not be able to complete developing FDA-authorized culture-based tests for resistance to a specific antibiotic until after the antibiotic is commercially available. The result is that the development of such culture-based tests may be generally delayed even after the new antibiotic is approved by FDA. This delay could affect the ability of clinicians to treat patients. For example, according to an expert, such a delay could lead to underuse of a newly available antibiotic, among other things, because a clinician may not be willing to prescribe the antibiotic without test results to guide treatment.

In addition to antibiotic developers waiting until FDA approves an antibiotic before a breakpoint is finalized, there are technical hurdles in developing a test for some new antibiotics, according to FDA officials. For

---

[73] According to FDA, different types of tests generally have different times for development after FDA approval of a new antibiotic. For example, some automated tests are the most delayed in being submitted for FDA clearance because of the sophistication of such tests. FDA officials told us one exception is for a type of test called "disk diffusion tests" that are evaluated as part of the drug approval process for new antibiotics. Such tests are often cleared within weeks of drug approval.

[74] Culture-based tests examine how well bacteria grow in the presence of an antibiotic. Another type of resistance test is a genotypic test, which directly detects genes that can make bacteria resistant. We focus on culture-based tests.

[75] One test manufacturer stated that there is a risk that the new antibiotic will ultimately not be approved by FDA. Thus, test manufacturers may wait for more certainty regarding the approval of a new antibiotic before committing resources to developing a test for it.

example, it may be challenging for certain automated test manufacturers to address unique growth properties of certain bacteria in the presence of specific antibiotics or combinations of antibiotics.

How Breakpoints Are Used to Interpret Tests

According to officials from the Food and Drug Administration (FDA), breakpoints, also referred to as “susceptibility test interpretive criteria,” are used to define susceptibility and resistance to antibiotics to help guide patient care. Culture-based tests rely on breakpoints to provide a determination of resistance to clinicians. In the United States, breakpoints (based on clinical or microbiological data) are established by standards- development organizations such as the Clinical and Laboratory Standards Institute (CLSI) and FDA.

One example of how breakpoints are used involves the Kirby-Bauer disk diffusion test. This test is conducted by spreading bacteria on a laboratory agar plate containing bacterial nutrients, and then placing paper disks containing a known amount of antibiotics on the “lawn” of bacteria. Plates are observed after overnight incubation to determine the extent of bacterial growth. Closer to the disk, there is a higher concentration of antibiotic, and the concentration declines with distance. Around most disks, there is a “zone of inhibition,” where the concentration of antibiotic is too high for bacteria to grow. After allowing the bacteria to grow for a defined period of time, the diameter of the zone of inhibition is measured in millimeters.

Procedure for Assessing Antibiotic Resistance Using Breakpoints

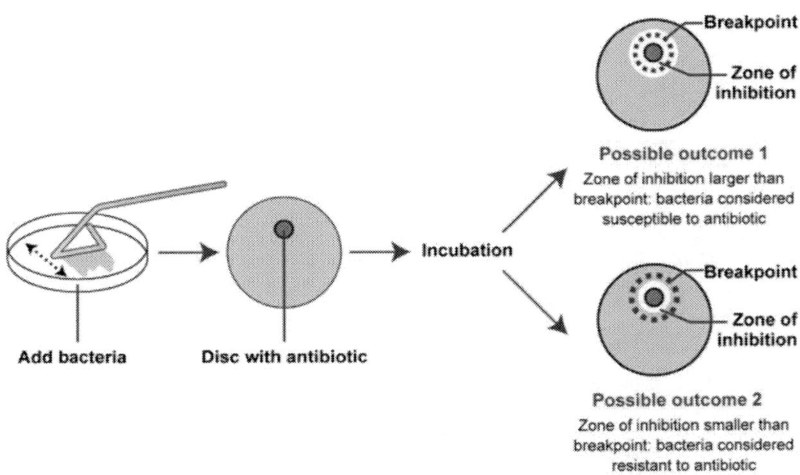

Source: GAO analysis of literature and technical documents. | GAO-20-341.

If the diameter is larger than or equal to the breakpoint, then the strain of bacteria is considered susceptible to the antibiotic, suggesting that the antibiotic can be used to treat infections caused by that strain. If the diameter is smaller than the breakpoint, then the strain is considered resistant, suggesting that the antibiotic should not be used. According to FDA, in most cases, there is a range of "intermediate" or "susceptible dose-dependent" diameters for which treatment might be effective.

Other types of culture-based diagnostic tests for resistance have analogous breakpoints for interpreting the test. For example, the minimum inhibitory concentration—the lowest concentration of an antibiotic that prevents growth of bacteria—can be compared to a breakpoint to establish whether the bacteria are considered resistant.

Source: GAO summary of interviews and documents from FDA and others. | GAO-20-341.

According to a test manufacturer, these hurdles include the need for additional studies, and such studies may not be straightforward because of the need to determine what clinical data FDA requires.[76] In addition, in the case of automated tests, a representative from a test manufacturer association told us the software used to run and interpret a new test needs to be revised, which can be time consuming.

The delay between approval of an antibiotic and the availability of a test for resistance could result in suboptimal treatment and increase burdens on the health care system. For example, one expert stated that during this delay, laboratories need to create or modify tests and then validate those tests instead of using a FDA-authorized test, which increases the time required and places demands on facility personnel and budgets. This expert added that to conduct validation studies, the laboratories need a variety of samples for testing, called "isolates," which may not be available. A second expert said that the delay leads to both overuse and underuse of the new antibiotic: in the absence of a test, some clinicians will prescribe the antibiotic when it may be inappropriate, leading to overuse;

---

[76] FDA officials told us FDA provides guidance documents that can help address this issue. These include Department of Health and Human Services, Food and Drug Administration, *Class II Special Controls Guidance Document: Antimicrobial Susceptibility Test (AST) Systems* (Aug. 28, 2009) and Department of Health and Human Services, Food and Drug Administration, *Requests for Feedback and Meetings for Medical Device Submissions: The Q-Submission Program, Guidance for Industry and Food and Drug Administration Staff* (May 7, 2019).

some other clinicians refrain from prescribing the antibiotic, even if appropriate, leading to underuse.

To help address this delay, FDA has created a process known as coordinated development, whereby test manufacturers can submit a coordinated development plan to FDA describing the test manufacturer's intent to coordinate with the antibiotic manufacturer. The plan is submitted prior to, or shortly after, submission of an application to market a new drug.[77]

Under the coordinated development program, FDA shares breakpoint information from the antibiotic manufacturer with a prospective test manufacturer.[78] It then reviews the test application at the same time as the antibiotic application and takes other steps to facilitate more timely clearance of the test. FDA officials told us this process has significantly reduced the delay between approval of the antibiotic and clearance of the test.[79]

Another FDA step to help test manufacturers speed development of tests is the establishment, in collaboration with CDC, of a centralized repository of bacterial strains with well-characterized antibiotic resistance profiles. These strains are available to test manufacturers and others to help them design, validate, and evaluate tests by checking that they give the correct results for bacteria whose profile of antibiotic resistance is known.[80] Finally, FDA officials also said that they offer pre-submission

---

[77] Department of Health and Human Services, Food and Drug Administration, *Coordinated Development of Antimicrobial Drugs and Antimicrobial Susceptibility Test Devices: Guidance for Industry and Food and Drug Administration Staff; Availability*, 84 Fed. Reg. 1152 (Feb. 1, 2019). FDA guidance pertaining to the development of antimicrobial drugs may apply to drugs that inhibit or destroy bacteria or fungi. See 21 U.S.C. § 321(jj); 42 U.S.C. § 247d-5(k).

[78] FDA officials told us this information can only be shared if the drug and test manufacturers permit it.

[79] For specific drugs approved since coordinated development was implemented, FDA provided some examples of the time from drug approval to device clearance for those tests that used coordinated development versus those that did not. While we did not independently confirm the times that resulted from the coordinated development process, examples of tests FDA provided us that used coordinated development had shorter times from drug approval until test clearance compared to tests that did not.

[80] See Food and Drug Administration, "The CDC and FDA Antimicrobial Resistance Isolate Bank," accessed January 21, 2020, https://www.fda.gov/MedicalDevices/Products andMedicalProcedures/InVitroDiagnostics/u cm454677.htm. CDC stated it is responsible

advice, whereby a test manufacturer can ask for initial guidance on the design of clinical studies for their tests.

FDA Has Taken Steps to Improve Breakpoint Recognition

In the United States, breakpoints are established and updated by organizations such as the Clinical and Laboratory Standards Institute (CLSI).[81] After CLSI establishes a breakpoint, FDA may review and recognize the breakpoint, according to FDA officials. Test manufacturers rely on breakpoints recognized by FDA to support marketing authorization of their tests.[82]

An expert who works for CLSI identified more than 50 breakpoints that have not been recognized by FDA, and for which CLSI considers FDA recognition important in order to help make FDA-authorized tests available.[83] Experts, including one from our meeting, cited the following examples of breakpoints needing recognition:

---

for the collection, maintenance, and restocking of these specimen banks. FDA noted that the day-to-day operation and the vast majority of the financial support is provided by CDC.

[81] The 21st Century Cures Act authorized FDA to recognize breakpoints developed by a nationally or internationally recognized standard development organization if the standard development organization meets specified criteria. 21st Century Cures Act, Pub. L. No. 114-255, §3044, 130 Stat. 1115-16 (Dec. 13, 2016), codified at 21 U.S.C. § 360a-2.
According to FDA, CLSI is the only organization for which FDA has made such a determination. Additionally, FDA also recognizes breakpoints that are different from CLSI, due to basic disagreements in scientific rationales for those breakpoints, among other reasons. One expert at our meeting told us that a common reason for FDA not recognizing a breakpoint is that the breakpoint may be associated with an unapproved, or "off-label," use of an antibiotic, such as use of the antibiotic at an unapproved dose. Once FDA approves a drug, clinicians sometimes prescribe the drug for an unapproved use when they judge that it is medically appropriate for their patient. However, our report is not only about unapproved uses. FDA-recognized breakpoints are available online as part of the 21st Century Cures Act requirements. See Food and Drug Administration, "Antibacterial Susceptibility Test interpretive Criteria," last accessed February 28, 2020, https://www.fda.gov/Drugs/DevelopmentApprovalProcess/DevelopmentResources/ucm57 5163.htm.

[82] Genotypic tests do not rely on breakpoints and can still be cleared by FDA in the absence of breakpoints.
FDA told us when FDA does not recognize a breakpoint, it may still identify a separate breakpoint, or interpretive criteria, appropriate for use by a test manufacturer in the development of its tests, under section 511A(b)(2)(B), 511A(c)(1) and (c)(2) of the Federal Food, Drug, and Cosmetic Act, codified at 21 U.S.C. §§ 360a-2(b)(2)(B), 360a- 2(c)(1), and 360a-2(c)(2), but FDA has not yet done so.

[83] While we did not independently verify the claim of importance, we confirmed that there are over 50 breakpoint discrepancies between a list provided by an expert working with CLSI in October 2018, and the breakpoints listed on the FDA website. See Food and Drug

- CDC recommends a dual therapy of antibiotics—azithromycin and ceftriaxone—to be taken together to treat gonorrhea.[84] However, FDA does not recognize any azithromycin breakpoints for *N. gonorrhoeae*, which an expert from our meeting told us could be a barrier to developing FDA-authorized culture-based tests for *N. gonorrhoeae* resistance to the recommended dual therapy.[85]
- Colistin is an antibiotic used in hospitals because of its efficacy against carbapenem-resistant bacteria, according to one manufacturer of a test for colistin resistance. This manufacturer markets its test in many countries but not in the United States, because FDA does not recognize colistin breakpoints.[86]

FDA has taken some steps to address unrecognized breakpoints, which are a potential barrier to developing some tests for antibiotic resistance.[87] For example, FDA officials told us that the agency conducts regular internal reviews of breakpoints. According to FDA officials, the agency

---

Administration, "Antibacterial Susceptibility Test Interpretive Criteria." CLSI does not consider all 50 breakpoints to be of the same priority, according to the expert. We did not confirm the list provided by the expert against CLSI documentation.

[84] According to CDC officials, CDC is currently in the process of updating their treatment guidelines for gonorrhea. The earlier recommendation for dual therapy for gonorrhea with azithromycin and ceftriaxone was based on the theoretical benefit of providing a shield (azithromycin) to protect ceftriaxone from increasing gonorrhea resistance. See K. A. Workowski and G. A. Bolan, "Sexually Transmitted Treatment Guidelines, 2015," *Morbidity and Mortality Weekly Report Recommendations and Reports,* vol. 64, no.3 (2015). Given increasing resistance to azithromycin now found in numerous pathogens and the continued importance of applying antibiotic stewardship principles to treatment decisions that impact the public's health, CDC is currently reviewing optimal gonorrhea treatment options for incorporation into forthcoming sexually-transmitted disease treatment guidelines.

[85] There are other challenges with making a test for resistant gonorrhea, including challenges in culturing the bacteria. However, without FDA-recognized breakpoints, surmounting those other challenges may not result in an FDA-authorized diagnostic test.

FDA officials told us they received a comment requesting recognition of this breakpoint in June 12, 2019.

[86] According to FDA, it published a notice on June 10, 2019, concluding that the scientific data available do not support recognizing breakpoints for colistin at this time.

[87] FDA officials told us that the biggest challenge that they see to recognizing breakpoints is the lack of necessary data to support breakpoint changes. FDA also told us it may not identify or recognize updated breakpoints unless there are data to support a scientific finding that breakpoints should be changed, under section 511A(a)(3) and (c)(2) of the Federal Food, Drug, and Cosmetic Act, codified at 21 U.S.C. §§ 360a-2(a)(3) and (c)(2).

reviewed the 2019 CLSI breakpoint standards and updated FDA's website with changes to recognized breakpoints as of June 2019. FDA has been posting such updates since December 13, 2017.[88]

FDA also accepts public comments requesting the recognition of new breakpoints, according to agency officials.[89] However, we found there was some confusion between CLSI officials and experts and the FDA involving the number of comments FDA could review each year, which FDA later clarified on its website. One expert at our meeting later told us that CLSI adjusts its process for submitting comments based in part on their understanding of FDA's communication. This expert added that FDA making a public commitment to a specific number of comments they would review would help CLSI improve its planning.

FDA officials told us there is no legal requirement for FDA to communicate the number of comments the agency can review, but that in previously published notices of opportunities for public comments, there was nothing that indicated there would be limits. However, after we informed FDA officials of concerns by experts regarding the number of comments FDA could review, FDA updated their webpage to clarify that they will review all submitted comments.[90]

---

[88] Food and Drug Administration, "Notice of Updates," accessed May 29, 2019, https://www.fda.gov/drugs/development-resources/notice-updates. These updates may reflect FDA's recognition of revised CLSI breakpoints or may reflect other types of changes to FDA's listed breakpoints, such as initial breakpoints identified by FDA at the time that a new drug is approved. FDA officials told us FDA is required by the 21st Century Cures Act to evaluate any appropriate new or updated breakpoints established by a standards development organization, as appropriate, every 6 months. Pub. L. No. 114- 255, § 3044(a), 130 Stat. 1114, codified at 21 U.S.C. § 360a–2(c)(1)(A).

[89] FDA reviews recommendations made in public comments under section 511A(c)(2) and (c)(3) of the Federal Food, Drug, and Cosmetic Act, codified at 21 U.S.C. §§ 360a-2(c)(2) and (c)(3).

[90] FDA officials told us they updated their breakpoint webpage to include the statement, "FDA will review all substantive submissions to support updating of susceptibility test interpretive criteria," to ensure clarity about FDA's commitment to reviewing all comments to the public docket. See Food and Drug Administration, "FDA-Recognized Antimicrobial Susceptibility Test Interpretive Criteria," accessed January 22, 2020, https://www.fda.gov/drugs/development-resources/fda-recognized-antimicrobial-susceptibility-test-interpretive-criteria.

FDA Has Taken Limited Steps to Monitor Use of Updated Breakpoints

FDA has taken limited steps to monitor whether FDA-authorized tests are using new breakpoints after these breakpoints are updated and accepted by FDA. Because bacteria can develop increasing resistance to antibiotics, it is sometimes important to change the breakpoints used for determining whether or not bacteria are resistant to a given antibiotic.

Using tests with out-of-date breakpoints could result in misidentifying a resistant infection as non-resistant, which can lead to treating a patient with an ineffective antibiotic and the further spread of the infection. FDA officials told us the agency has taken limited steps to monitor the status of breakpoint updates, and that out-of-date breakpoints being used in tests should be a rare occurrence.[91]

In contrast, a CDC official told us that keeping tests updated is a significant concern. This official cited the example of carbapenem-resistant Enterobacteriaceae infection, which triggers specific procedures to limit the spread of these bacteria. If the test breakpoint is out of date, the infection may not be detected in a timely manner, and the pathogen could spread broadly as a result. A recent study looking at hypothetical scenarios in one U.S. county estimated that a 32-month delay in updating tests to match CLSI breakpoints for carbapenem-resistant Enterobacteriaceae would have resulted in an average of almost 2,000 additional carriers of these bacteria county-wide.[92] Additionally, an expert told us that use of

---

[91] For example, FDA officials told us of its implementation of "change protocols" in 2019 for certain tests that were first implemented in 2019. These protocols permit device manufacturers, under certain conditions, to update test breakpoints frequently for cleared tests without the need to submit a new 510(k) for FDA review. FDA told us it expects this protocol will significantly reduce the time frame for updating device breakpoints when new breakpoints are published, while maintaining test safety and efficacy. FDA officials explained that this protocol will encourage manufacturers to incorporate up-to-date breakpoints in their tests by not requiring a regulatory submission in most, if not all, instances. However, FDA did not explain how this relates to FDA's monitoring, or addressing of existing out-of-date breakpoints.

[92] S. M. Bartsch et al., "Impact of Delays between Clinical and Laboratory Standards Institute and Food and Drug Administration Revisions of Interpretive Criteria for Carbapenem-Resistant Enterobacteriaceae," *Journal of Clinical Microbiology* vol. 54, no. 11 (2016): pp. 2757-2762. The countywide average is the average number of additional carriers among facility types—acute care hospitals, long-term acute care facilities, and free-standing nursing homes. The reported countywide average from a 2.5-year delay was 1,821.0 more

out-of-date breakpoints could lead to improper patient care, improper surveillance reporting, and slower detection of emerging resistance.[93] However, the true impact of this issue is challenging to discern (see text box).

The Extent of Any Negative Effects of Out-of-Date Breakpoints on Public Health Is Unclear

Experts and agency officials voiced a range of opinions on the public health effects of tests with out-of-date breakpoints. For example, one Centers for Disease Control and Prevention (CDC) official told us that despite the lack of breakpoint updates, cases of a type of carbapenem-resistant Enterobacteriaceae were likely ultimately caught by hospitals because a second test was used by all but a small number of hospitals. One expert stated that how quickly test breakpoints are updated is less important when deciding what test to adopt than other factors, such as ease of use. However, another expert noted that laboratories addressing emerging threats may feel the need to use non-Food and Drug Administration (FDA) cleared tests, because they are aware that FDA-cleared tests may not be updated as quickly as needed. Test updates may be an issue for smaller laboratories, which do not have dedicated personnel keeping track of breakpoint revisions, Department of Veterans Affairs officials told us.

Source: GAO summary of interviews with CDC and others. | GAO-20-341.

FDA officials told us that because manufacturers are strongly motivated to keep their tests current, only a few tests have out-of-date breakpoints.[94]

---

carbapenem-resistant Enterobacteriaceae carriers (95 percent confidence interval: 1,009.6 to 2,632.4). FDA officials noted that this publication predated enactment of the 21st Century Cures Act, which provides for a more streamlined process for incorporating up-to-date information into tests. However, FDA did not discuss how the Act might affect the outcomes discussed in the study.

[93] FDA officials told us genomic tests are important for identifying bacteria such as carbapenem-resistant Enterobacteriaceae and that some genomic tests are increasingly used. However, they did not address how genomic tests may affect the use of existing tests with out-of-date breakpoints.

[94] FDA told us that it considers this to be the case mainly because manufacturers want to make sure their devices are updated. However, FDA did not provide evidence for this assertion, and one test manufacturer told us they question this assessment.

FDA recommends, but does not require, that test manufacturers update their labeling to conform to new, publicly available breakpoints within 90 days. Department of Health and Human Services, Food and Drug Administration, *Guidance for Industry, Updating Labeling for Susceptibility Test Information in Systemic Antibacterial Drug Products and*

However, the only confirmation FDA officials offered for this statement was to mention an unofficial internal survey of FDA's database of existing tests, conducted in March 2019, which concluded that all FDA-authorized tests had implemented breakpoint updates made since December 13, 2017.[95] They said this survey is not conducted regularly. They also stated that it is possible that some tests have not been updated to reflect breakpoint updates made prior to December 13, 2017, but that FDA is unaware of any such tests that also pose a public health threat.[96]

To assess the extent to which there are FDA-authorized tests using out-of-date breakpoints, we spoke with experts and stakeholders and reviewed studies they identified. We identified several FDA-authorized tests with breakpoints that were changed nearly a decade ago. Some of these tests could be used for diagnosing infection with carbapenem- resistant Enterobacteriaceae, which CDC identified as an urgent threat. One manufacturer told us that one of their tests has not been updated with new breakpoints nearly 10 years after a breakpoint revision. FDA officials acknowledged it is possible some FDA-authorized tests might continue to rely on outdated breakpoints.[97] Further, in 2019, a scientific article listed four different test manufacturers offering tests that have not been fully updated to reflect revised breakpoints, including some affecting antibiotics

---

Antimicrobial Susceptibility Testing Devices; *Availability,* 74 Fed. Reg. 31740 (July 2, 2009).

[95] FDA restricted its survey to those breakpoint updates listed on the FDA's breakpoint web site and excluded newly approved drugs, since it does not consider breakpoints for new drugs to be breakpoint changes.

[96] FDA officials noted that no actual harm has to occur to be considered a public health threat. They told us they assess public health threats using a risk-assessment framework, in which FDA analyzes the frequency or severity of an inaccurate result, among other things. Additional information can be found at Department of Health and Human Services, Food and Drug Administration, *Factors to Consider Regarding Benefit-Risk in Medical Device Product Availability, Compliance, and Enforcement Decisions: Guidance for Industry and Food and Drug Administration Staff* (Rockville, Md.: Dec. 27, 2016). FDA told us that some tests using out-of-date breakpoints may not pose a public health concern due to various mitigating factors. FDA noted that identifying cause of concern is complicated and requires deep analysis.

[97] FDA officials told us that the tests being asked about provide information for the user to interpret based on criteria other than FDA-recognized breakpoints. However, using this approach typically requires additional studies by the test user.

for some types of carbapenem-resistant Enterobacteriaceae.[98] Finally, CDC officials told us they asked hospital laboratories in a survey for 2017 and 2018 if they had updated their tests to reflect revisions in breakpoints for carbapenem-resistant Enterobacteriaceae that were implemented in 2010. According to CDC, nearly 1,000 of over 5,000 responding hospital laboratories had not implemented the revised breakpoints, and, of these, over 85 percent were using FDA-authorized tests.[99]

One CDC official stated that there is significant concern for patient safety associated with out-of-date breakpoints, and another said that there are few justifications for failing to update the tests after 8 years. FDA officials told us they have not received reports of suspected device-associated deaths, serious injuries, or malfunctions that are specific to out-of-date carbapenem-resistant Enterobacteriaceae breakpoints in FDA-authorized tests using such breakpoints. The officials added that it is possible to detect carbapenem-resistant Enterobacteriaceae under certain situations, even if the test had an out-of-date breakpoint for a given antibiotic against these bacteria.

However, FDA does not know the actual negative effect, if any, of out-of- date breakpoints because it does not know how many FDA-authorized tests rely on such breakpoints. Since December 2017, FDA has conducted one unofficial survey of tests to assess breakpoint updates that was limited in scope and is not a regular event. Other than that, FDA is relying on market incentives to drive manufacturers to make sure their devices are updated.[100]

According to FDA and others, the extent of the problem is not clear. However, PACCARB identified updating test breakpoints as an important

---

[98] R. M. Humphries, A. N. Abbott, and J. A Hindler, "Understanding and Addressing CLSI Breakpoint Revisions–A Primer for Clinical Laboratories," *Journal of Clinical Microbiology,* vol. 57, no. 6 (2019): pp. e00203-19.

[99] In some cases, diagnostic test users could choose not to update their test(s). For example, VA officials note that there may be good reasons why test users may not want updated breakpoints, such as different antibiotic resistance patterns in their locale.

[100] FDA told us they also evaluate risk by review of medical devices reports, attendance at major scientific meetings, and participation in CLSI activities, among other activities. However, evaluation of risk is not the same as monitoring status.

issue in a 2017 report.[101] Additionally, one of the sub-objectives in the National Action Plan notes that rapid updating of breakpoints is essential to provide accurate information to guide appropriate drug treatment.[102] Finally, the Standards for Internal Control in the Federal Government directs management to establish and operate monitoring activities to monitor its internal control systems and evaluate the results.[103] In this case, monitoring and evaluation of the status of breakpoint updates in FDA-authorized tests could help FDA identify and address the National Action Plan sub-objective, as well as a strategic priority in the mission statement of its Center for Devices and Radiological Health: "FDA assures that patients and providers have timely and continued access to safe, effective and high-quality medical devices."[104]

FDA officials said they do not believe the issue is a significant problem, but the agency has also not regularly evaluated any effects of using tests for antibiotic resistance with out-of-date breakpoints.[105] FDA officials stated that there may be resource constraints to their ability to conduct regular monitoring and evaluation. By regularly monitoring and evaluating FDA-authorized tests, FDA would be better positioned to determine the extent of tests relying on out-of-date breakpoints and may be better positioned to provide assurance that patients and providers have timely access to safe and effective tests. Furthermore, by regular monitoring, FDA would be able to determine whether test manufacturers are updating breakpoints as needed, and help ensure that patient care and infection control efforts are effective.

---

[101] Presidential Advisory Council on Combating Antibiotic-Resistant Bacteria, *Recommendations for Incentivizing the Development of Vaccines, Diagnostics, and Therapeutics*. The relevant report section focused on microbiology laboratories' updating of technologies, in addition to breakpoints, which could fall under CMS actions. We focus on FDA given that the majority of tests in use are FDA-authorized tests, and our primary attention is on breakpoint updates.

[102] The White House, *National Action Plan,* sub-objective 1.1.5, p. 19.

[103] GAO-14-704G.

[104] Food and Drug Administration, *2018-2020 Strategic Priorities: Center for Devices and Radiological Health* (January 2018), p. 3.

[105] FDA officials told us they continue to use various postmarket surveillance tools to monitor device performance, detect device-related safety issues, and contribute to benefit- risk assessment of devices including these tests. However, they did not discuss how these tools enable FDA to proactively determine whether tests are not being updated in a timely manner.

Federal Efforts Have Not Fully Addressed Challenges to Developing New Treatments for Antibiotic-Resistant Infections

Experts, federal officials, and antibiotic developers have identified economic and other challenges to developing new antibiotics. Federal agencies, including HHS and DOD, have engaged in efforts to address some of the challenges; however, experts said these efforts are not sufficient and that additional federal incentives are needed to encourage the development of new antibiotics.

Economic and Other Challenges to Developing New Treatments Exist

Experts are concerned about a void in the discovery of new antibiotic classes and the current pipeline of antibiotics in development. According to The Pew Charitable Trusts, a nonprofit public policy organization that tracks the pipeline of antibiotics, no new classes of antibiotics approved for human use have been discovered since 1984.[106] In addition, experts are concerned that the number of antibiotics in clinical development is insufficient to meet the threat of antibiotic resistance. For example, according to The Pew Charitable Trusts, only 42 antibiotics were in clinical development globally—meaning clinical trials were being conducted to test their safety and efficacy in humans—as of June 2019, and only 24 of them targeted bacteria on CDC's or WHO's priority lists.[107] According to a recently published analysis, the authors found that the pipeline of antibiotics that target gram-negative bacteria is dominated by

[106] This discovery void is concerning because when bacteria become resistant to one type of antibiotic, it is often resistant to others within the same class. Drugs can be classified according to similarities in chemical structure, among other ways.

[107] The Pew Charitable Trusts, "Antibiotics Currently in Global Clinical Development," accessed June 4, 2019, https://www.pewtrusts.org/en/research-and-analysis/data-visualizations/2014/antibiotics-currently-in-clinical-development.

derivatives of existing classes of antibiotics and "does not sufficiently address the problem of extensively drug-resistant gram-negative bacteria."[108] Experts from our meeting, antibiotic developers, and federal officials we spoke with identified major economic challenges to developing new antibiotics that commonly result from a poor return on investment. Antibiotics, like other types of pharmaceutical drugs, require substantial investment and many years to bring a new drug to market. We previously reported that for a new drug, the entire R&D process, including basic research, drug discovery, clinical trials, and FDA review, can take up to 15 years, often accompanied by high costs.[109] We also previously reported that an important incentive for pharmaceutical R&D investment is the potential for high revenue typically associated with a large number of patients or high drug prices.[110] However, sales revenues for brand- name antibiotics are often low. For example, studies found that median annual sales for brand name antibiotics between 2011 and 2015 ranged from $24 million to $75 million, whereas annual sales for most new, brand-name oncology drugs were more than $500 million during the same period.[111] Antibiotic developers we interviewed and experts have identified multiple factors that can limit the profitability of new antibiotics when they reach the market, including:

---

[108] U. Theuretzbacher et al., "Analysis of the Clinical Antibacterial and Antituberculosis Pipeline," *The Lancet Infectious Diseases*, vol. 19, no. 2 (2019): pp. 40-50. Specifically, the authors found that the pipeline does not sufficiently address the bacteria *A. baumannii*, *P. aeruginosa,* and Enterobacteriaceae.

[109] GAO, *Drug Industry: Profits, Research and Development Spending, and Merger Acquisition Deals*, GAO-18-40 (Washington, D.C.: Nov. 17, 2017).
For example, one study estimated the average cost per new molecular compound that received FDA approval between 2005 and 2013 to be $1.4 billion. See J. A. DiMasi, H. G. Grabowski, and R. W. Hansen, "Innovation in the Pharmaceutical Industry: New Estimates of R&D Costs," *Journal of Health Economics*, vol. 47 (2016): pp. 20-33. Other studies suggest lower development costs. For example, another study estimated a median cost to develop cancer drugs of $0.6 billion. See V. Prasad and S. Mailankody, "Research and Development Spending to Bring Single Cancer Drug to Market and Revenues After Approval," *JAMA Internal Medicine*, vol. 177, no. 11 (2017): pp. 1,569-1,575.

[110] GAO-18-40.

[111] Duke Margolis Center for Health Policy, *Value-Based Strategies for Encouraging New Development of Antimicrobial Drugs,* and IMS Institute for Healthcare Informatics, *Medicines Use and Spending in the U.S. A Review of 2015 and Outlook to 2020* (Parsippany, N.J.: April 2016).
The Duke Margolis Center also reported that only five of the 16 new brand-name antimicrobials approved since 2000 generated annual sales above $100 million.

- Prices for new antibiotics are considered low compared to other life- saving drugs, such as cancer drugs, because they must compete with inexpensive generic antibiotics, which remain effective enough to influence pricing of new antibiotics.[112]
- Antibiotics are typically used for a short duration, unlike drugs for chronic diseases that patients use for many months or years.
- Antibiotic stewardship principles encourage the appropriate use of antibiotics to help prevent resistance. According to the Duke University Margolis Center for Health Policy, most novel treatments for antibiotic-resistant infections have a narrow set of patients for whom the treatment would be appropriate.[113]

Overview of Clinical Trials

Before approving a new drug, the Food and Drug Administration (FDA) generally requires a drug's developer to conduct clinical trials in humans to assess its safety and effectiveness against a specific disease or illness—called an "indication." Two types of trials that may be used are:

- Superiority trials, which aim to show that the drug being investigated is more effective than an existing drug.
- Non-inferiority trials, which aim to demonstrate that the difference between the effectiveness of the drug being investigated and an existing drug is small enough to show that the drug being studied is also effective.

Typically, there are three phases of clinical trials, with the sizes of the trials increasing with each phase. FDA generally prefers that when conducting clinical trials, developers demonstrate the effectiveness of a new drug by showing its impact on a clinical endpoint— a direct measure of how a patient feels, functions, or survives. FDA also accepts surrogate endpoints, which are laboratory measures or physical signs used as a substitute for a clinical endpoint that reasonably predict a clinical benefit.

Source: GAO and FDA. | GAO-20-341.

[112] Some experts believe the prices of new antibiotics do not reflect their public health value–that is, the value they provide not just to the patient, but also to society—because effective treatment prevents the antibiotic-resistant bacteria from spreading. See Duke Margolis Center, *Value-Based Strategies for Encouraging New Development of Antimicrobial Drugs.*

[113] Duke Margolis Center, *Value-Based Strategies for Encouraging New Development of Antimicrobial Drugs.*

As a result of the perceived poor return on investment, many large pharmaceutical companies have discontinued their antibiotic development in recent years. In 2018, according to The Pew Charitable Trusts and other published sources, four large pharmaceutical companies worldwide had antibiotics in clinical development globally compared to 1990, when 18 were involved in antibiotic R&D.[114] Two antibiotic companies declared bankruptcy in 2019; in the case of one, the company filed for bankruptcy only 10 months after its antibiotic, which targets resistant bacteria, received FDA approval.[115] The majority of antibiotics in the development pipeline are being developed by smaller companies that do not have other drugs on the market to help cover their R&D costs. However, representatives from three small antibiotic developers we spoke with noted that their field is struggling because it is difficult to raise funds from private investors due to the low return on investment potential.

Further complicating these economic challenges, federal officials and antibiotic developers we spoke with also identified challenges in conducting clinical trials for antibiotics, which can make it difficult to meet FDA's regulatory requirements for approval.[116] This is particularly true for antibiotics that would treat antibiotic-resistant infections. Specifically, they noted the following three challenges:

- *Enrolling patients in clinical trials.* Identifying and enrolling patients with bacterial infections into certain clinical trials prior to initiating treatment can be difficult due to a lack of available rapid diagnostic tests to identify the type of infection and the urgent

---

[114] The Pew Charitable Trusts, "Tracking the Global Pipeline of Antibiotics in Development," accessed November 29, 2019, https://www.pewtrusts.org/en/research-and-analysis/issue-briefs/2019/03/tracking-the-global-pipeline-of-antibiotics-in-development; and P. Sharma and A. Towse, Office of Health Economics, *New Drugs to Tackle Antimicrobial Resistance: Analysis of EU Policy Options* (London, England: April2011).

[115] A third company, which gained FDA approval for an antibiotic that treats complicated intra-abdominal infections in 2018, announced in 2019 that it would reduce its workforce and cease research functions to focus its financial resources on commercializing its approved antibiotic. In March 2020, the company announced it would be acquired by another company.

[116] Before approving a new drug, FDA generally requires the developer to conduct clinical trials in humans to assess its safety and effectiveness against a specific disease or illness—called an "indication."

need to begin treatment immediately for acute infections. According to FDA officials, this is problematic for clinical trials because any prior treatment could obscure the true efficacy of the drug under investigation. Recognizing this often unavoidable issue, FDA has issued guidance giving antibiotic developers additional, but limited, flexibility with their clinical trial protocols in certain cases.[117]

In addition, certain types of antibiotic-resistant infections are rare and, therefore, antibiotic developers and federal officials told us it can be difficult to find patients to enroll in clinical trials to test antibiotics that target resistant bacteria. FDA officials told us that, for some types of bacterial infections, only 5 to 10 percent of patients have an infection caused by a resistant bacterial strain.

- *Demonstrating superiority of a new antibiotic.* Two antibiotic developers told us that, for most antibiotics, it is difficult to conduct superiority clinical trials and more feasible to conduct non-inferiority trials, because the latter allows for smaller enrollment.[118] (See side bar for an explanation of clinical trial types.) They told us that the inability to demonstrate their drug's superiority limits their ability to market the drug, because it can be difficult to convince purchasers (e.g., hospitals) to choose the newly approved antibiotic over existing antibiotics, especially when the new antibiotic is significantly more expensive.
- *Gaining approval for multiple indications.* FDA generally approves drugs for a specific indication; therefore, antibiotic

---

[117] For example, in its guidance about developing drugs for complicated urinary tract infections, FDA states that up to 25 percent of patients in a clinical trial may have received another type of drug before receiving the drug under investigation. Department of Health and Human Services, Food and Drug Administration, *Complicated Urinary Tract Infections: Developing Drugs for Treatment: Guidance for Industry* (Silver Spring, Md.: June 2018).

[118] According to literature we reviewed, ethical clinical trial design for serious and life-threatening infections requires trials to compare the drug being studied to an existing treatment for that infection, if it is known to be safe and efficacious. However, it is unlikely that superiority of a new drug would be observed when the bacteria are susceptible to both drugs. See H. Boucher et al., "White Paper: Developing Antimicrobial Drugs for Resistant Pathogens, Narrow-Spectrum Indications, and Unmet Needs," *The Journal of Infectious Diseases*, vol. 216, no. 2 (2017): pp. 228-236.

developers told us they tend to design their clinical trials around common infection types, largely because of the relative ease of enrolling patients. However, some antibiotics can treat infections in multiple parts of the body, which may not have been studied in a clinical trial.[119] While providers are able to prescribe drugs for "off-label" use—that is, for a condition or patient population for which the drug has not been approved—they may lack information on the safety and efficacy of the drug for such use. In addition, such off-label use may not be reimbursed by the patient's insurance.

Nontraditional Products in Development

According to The Pew Charitable Trusts, there were 29 nontraditional antibacterial products in clinical development for the U.S. market in June 2019. Among the 29 products in the pipeline, nine were antibodies, seven were vaccines, seven were live biotherapeutic products, and six were other types of products. No bacteriophages were in clinical development. More than half of these products are for the treatment of *Clostridioides difficile* or *Staphylococcus aureus* infections.

Source: The Pew Charitable Trusts. | GAO-20-341.

Experts, antibiotic developers, and federal officials also said it is scientifically challenging to develop new antibiotics that can overcome existing mechanisms of resistance. One expert at our meeting explained that it is necessary to develop an antibiotic that works differently than existing antibiotics so that bacteria are not resistant to it. In particular, experts and federal officials have noted that it is challenging to develop antibiotics that can kill certain types of bacteria, called gram-negative bacteria, largely due to their double membrane that makes it difficult for

---

[119] In 2017 guidance, FDA stated that it would allow antibiotic developers flexibility to include, in certain superiority trials, patients with infections in multiple body sites caused by the same bacteria. However, superiority trials are often difficult to conduct for antibiotics, as described above. Department of Health and Human Services, Food and Drug Administration, *Antibacterial Therapies for Patients with an Unmet Medical Need for the Treatment of Serious Bacterial Diseases; Guidance for Industry; Availability*, 82 Fed. Reg. 35973 (Aug. 2, 2017).

antibiotics to enter the bacterial cell, and to pumps that can remove the drug once it enters. Three antibiotic developers we spoke to explained that as bacteria continue to evolve new ways to resist antibiotics, it is difficult for scientists to keep pace by developing new treatments that can overcome those mechanisms. In addition, experts noted that scientists have already discovered most of the antibiotics from known sources, such as soil. As a result, scientists are now exploring new sources of chemicals with antibiotic properties, such as insects.[120]

As the rate of antibiotic discovery has slowed, scientists have also begun to explore alternatives to traditional antibiotics—which we call "nontraditional products" in this report.[121] Many types of nontraditional products are currently being researched and developed to treat antibiotic-resistant infections, including, among others, live biotherapeutic products, antibodies, and bacteriophages.[122] For example, one type of nontraditional product in use for the treatment of recurrent *C. difficile*- associated disease—which causes diarrhea, abdominal cramps, and an estimated 15,000 deaths in the United States each year, according to CDC—is fecal microbiota for transplantation, more commonly known as fecal transplants.[123] (See text box.) However, scientists and companies researching and developing certain types of nontraditional products face development challenges.

---

[120] For example, scientists have recently discovered that chemicals within insects and komodo dragons have antibiotic properties, which could potentially be used as the basis for developing synthetic antibiotics in the future.

[121] According to NIAID, nontraditional products are antibacterial agents or approaches that differ in mechanism from traditional small-molecule agents that kill bacteria or inhibit their growth.

[122] Live biotherapeutic products are products that contain live organisms, such as bacteria. Antibodies are proteins naturally produced by the body's immune system to help remove potentially harmful pathogens. Antibodies can be harvested and used as medicines. Bacteriophages are viruses that can kill bacteria.

[123] CDC considers *C. difficile* to be an urgent threat. While *C. difficile* is not usually resistant to antibiotics, it is caused by the same factors that drive antibiotic resistance— antibiotic use and the spread of germs.

Fecal Transplants

The goal of a fecal transplant—which involves collecting stool from healthy donors and transferring it to patients via enema, oral capsule, or another modality—is to restore a healthy gut microbiome for recipients. According to the National Institutes of Health (NIH), multiple research studies have indicated that these transplants are effective, but their long-term safety has not been established. Questions remain about the Food and Drug Administration's (FDA) policy regarding stool banks that collect, prepare, and distribute fecal transplant products. FDA issued guidance in 2013 indicating its intention to exercise enforcement discretion regarding Investigational New Drug requirements for the use of fecal transplants to treat *Clostridioides difficile* infections, provided that the treating physician obtained adequate consent from the patient or his or her legally authorized representative. In other words, FDA's guidance indicated it would not require fecal transplant products to satisfy the Investigational New Drug requirements— which refer to the requirements for FDA's approval before beginning clinical trials to test a product on humans. [FDA, *Enforcement Policy Regarding Investigational New Drug Requirements for Use of Fecal Microbiota for Transplantation To Treat Clostridium difficile Infection Not Responsive to Standard Therapies; Guidance for Industry; Availability*, 78 Fed. Reg. 42965 (Jul. 18, 2013).] However, FDA later issued draft guidance in 2016 stating that FDA did not intend to extend enforcement discretion with respect to the Investigational New Drug requirements applicable to stool banks distributing fecal products. [FDA, *Enforcement Policy Regarding Investigational New Drug Requirements for Use of Fecal Microbiota for Transplantation To Treat Clostridium difficile Infection Not Responsive to Standard Therapies; Draft Guidance for Industry; Availability*, 81 Fed. Reg. 10632 (Mar. 1, 2016).] FDA has not finalized the 2016 draft guidance, which leaves the final guidance from 2013 as the current policy.

According to FDA, the agency received many comments from patients and industry groups in response to the 2016 draft guidance expressing concern about the effect that the requirement for clinical trials would have on access to these products. In March 2019, FDA officials told us they were still reviewing comments to the 2016 draft guidance and were unable to say whether or not it would be finalized. In November 2019, FDA held a public hearing to obtain further input on the use of fecal transplants to treat *C. difficile* infection not responsive to standard therapies and to better understand the effect of FDA's enforcement policy on product development.

Source: GAO analysis of information from NIH and FDA. | GAO-20-341.

For example, according to a paper written by BARDA officials and others, certain types of nontraditional products target only one or a few types of bacteria, which makes enrollment of patients in clinical trials difficult and potentially cost-prohibitive.[124] The authors also stated that additional research is needed to evaluate side effects and measure the efficacy of some types of nontraditional products. According to another published paper, more than half of the nontraditional products in development are intended to be used concurrently with a traditional antibiotic, and it can be difficult to demonstrate the additional clinical benefit of adjunctive therapies in clinical trials.[125] The authors also noted that additional clinical trial endpoints still need to be developed and validated for such nontraditional products.

Federal Agencies Have Made Some Progress toward Addressing Treatment Development Challenges

Multiple federal agencies have supported the development of new antibiotic treatments, including providing funding for antibiotic R&D, issuing guidance related to antibiotic clinical trials, and implementing Medicare payment mechanisms. Agencies have made available both "push" incentives, which directly support antibiotic R&D, and "pull" incentives, which offer financial benefit, either directly or indirectly, to developers of successful antibiotics after they reach the market.

Federal Funding for Antibiotic R&D

Several federal agencies award grants or contracts, create public-private partnerships, or use other approaches to provide researchers the funding for R&D of new treatments for antibiotic-resistant infections (see Table 3). This type of pre- market R&D support is considered a "push incentive."

---

[124] B. N. Tse et al., "Challenges and Opportunities of Nontraditional Approaches to Treating Bacterial Infections," *Clinical Infectious Diseases*, vol. 65, no. 3 (2017): pp. 495-500.

[125] U. Theuretzbacher et al., "The Global Preclinical Antibacterial Pipeline," *Nature Reviews Microbiology* (Nov. 19, 2019).

Table 3. Examples of federal agencies' funding of research and development for new treatments for antibiotic-resistant infections

Agency	Examples of efforts
Department of Health and Human Services (HHS) Biomedical Advanced Research and Development Authority (BARDA)	BARDA officials told us BARDA has awarded $959 million in the form of grants, cooperative agreements, and contracts to developers of 24 antibiotic drugs and one nontraditional product since 2010.[a] As of September 2019, three of these antibiotics had been approved by the Food and Drug Administration for marketing.
	BARDA created public-private partnerships with four antibiotic development companies that are developing a total of seven antibiotic candidates among them, investing a total of nearly $403 million since 2013, according to BARDA officials.
	BARDA led the creation of the Combating Antibiotic-Resistant Bacteria Biopharmaceutical Accelerator (CARB-X) in 2016, an international public-private partnership that provides funds to support preclinical research (i.e., research prior to beginning testing in humans) and phase 1 clinical trials. BARDA has committed to provide up to $250 million of funding to CARB-X.[b] According to its 2019 annual report, CARB-X had funded 47 projects, totaling up to $133.5 million.[c]
Department of Health and Human Services (HHS)National Institutes of Health (NIH)	NIH officials estimated that the National Institute of Allergy and Infectious Diseases (NIAID) awarded $158 million in fiscal year 2017 and $148 million in fiscal year 2018 in grants, contracts, and other funding mechanisms for the study of treatments for antibiotic-resistant infections. These figures do not include funding of basic research that can lead to the development of new treatments or the improvement of existing treatments.
	NIH supports CARB-X by providing research services to support awardees and holding key governance roles.

Table 3. (Continued)

Agency	Examples of efforts
Department of Defense (DOD) Defense Threat Reduction Agency, U.S. Army Medical Research and Materiel Command, and others	DOD funds and conducts research on treatments for antibiotic-resistant infections, with total awarded funding of about $271 million since 2012, according to DOD officials.[d] For example, the Defense Threat Reduction Agency has awarded 21 projects, totaling approximately $178 million, and the U.S. Army Medical Research and Materiel Command has awarded 50 projects, totaling $66.2 million.[e]

Source: GAO summary of information from HHS and DOD officials and agency documentation. | GAO-20-341.

[a]According to BARDA officials, funding for these awards is provided from the Public Health and Social Services Emergency Fund.

[b]According to HHS officials, BARDA's funding for CARB-X is provided from the Public Health and Social Services Emergency Fund. Other contributors to CARB-X include the Wellcome Trust—a global health charitable foundation, the governments of the United Kingdom and Germany, and the Bill and Melinda Gates Foundation.

[c]CARB-X requires grantees to share in the costs of research and development by contributing at least 30 percent of the cost of the project.

[d]According to DOD officials, they use grants, contracts, direct funding, and other transaction authority. Other Transaction Authority, in this case, is the term used by DOD to refer to legal acquisition instruments other than contracts, grants, or cooperative agreements, to carry out certain prototype projects. 10 U.S.C. § 2371b.

[e]According to DOD officials, the U.S. Army Medical Research and Materiel Command was redesignated as the U.S. Army Medical Research and Development Command in June 2019.

CARB-X's Portfolio

The Combating Antibiotic-Resistant Bacteria Biopharmaceutical Accelerator (CARB-X) had funded 47 projects, with the following in its portfolio as of July 2019:

- 12 antibiotics,
- 10 nontraditional therapies,
- 3 vaccines, and
- 5 diagnostic tests.

Among the products in the CARB-X portfolio, 12 would represent a new antibiotic class (if approved) and 14 target a novel molecular bacterial target. Awardees were based in six countries.

Source: CARB-X 2018-2019 Annual Report. | GAO-20-341.

See appendix III for additional examples of efforts to support antibiotic R&D by NIH and DOD.

Issued Guidance to Support Clinical Trials

FDA has implemented programs and issued guidance that help address some regulatory challenges and encourage antibiotic development. In 2012, through the Generating Antibiotic Incentives Now provisions of the Food and Drug Administration Safety and Innovation Act, Congress created the Qualified Infectious Disease Product (QIDP) designation.[126] Drugs that FDA designates as QIDPs, which include antibiotics and antifungals, may qualify for 5 years of additional exclusivity and fast-track or priority review designation during the FDA review process.[127] The additional exclusivity conferred to QIDP designees is a type of "pull incentive," because it offers the potential for enhanced financial gain after a drug receives FDA approval and reaches the market. According to FDA officials, as of September 2019, FDA had granted 192 QIDP designations, 24 of which it has approved for marketing.[128]

Also in response to the Generating Antibiotic Incentives Now Act, FDA released final guidance in August 2017 to streamline clinical development of antibiotics for patients with an unmet medical need—that is, those with a serious bacterial disease that has few or no treatment options.[129] FDA explains in this guidance that it may consider drugs for

---

[126] Pub. L. No. 112-144, tit. 8, §§ 801 et seq., 126 Stat. 993, 1077 (2012). The Generating Antibiotic Incentives Now Act defines a QIDP as an antibacterial or antifungal drug for human use intended to treat serious or life-threatening infections. Pub. L. No. 112-144, § 801, 126 Stat. 1077 (codified in pertinent part at 21 U.S.C. § 355f(g)). In this report, we refer to title 8 of Pub. L. No. 112-144 as the Generating Antibiotic Incentives Now Act.

[127] Pub. L. No. 112-144, §§ 801-803, 126 Stat. 1077-1079 (codified in pertinent part as amended at 21 U.S.C. §§ 355f (a), (d) (exclusivity extension and QIDP designation), 360n-1 (priority review), 356 (b) (fast track)).
In this report, our use of the term "exclusivity" refers to exclusive marketing rights granted by law for certain periods upon approval of a drug application, if certain requirements are met. QIDP exclusivity is granted as an extension to certain other exclusivity for which the applicant qualifies under the Federal Food, Drug, and Cosmetic Act.

[128] For example, FDA approved a new drug for the treatment of a specific type of highly drug-resistant tuberculosis in August 2019.

[129] Pub. L. No. 112-144, § 806, 126 Stat. 1082 (codified at 21 U.S.C. § 355 note). Food and Drug Administration, *Antibacterial Therapies for Patients with an Unmet Medical Need; Guidance for Industry.*

these patients that have higher risks than would be acceptable for a broad patient population and provides information on types of antibiotics that could be eligible for approval based on smaller, shorter, or fewer—as few as only one— clinical trials.

The 21st Century Cures Act required FDA to establish a Limited Population Pathway for Antibacterial and Antifungal Drugs (LPAD).[130] In June 2018, FDA issued draft LPAD guidance, as required by the Act.[131] Under LPAD, eligible products—which are drugs and biologics intended to treat a serious or life-threatening infection in a limited population of patients with unmet needs—may follow a streamlined development program, similar to the approaches described in its earlier unmet medical need guidance.[132] A biotechnology association noted in its public comments to the draft LPAD guidance the need for FDA to issue additional guidance to clarify its expectations for acceptable types of efficacy data when clinical trials are small and to clarify its interpretation of a "limited population of patients" for the purpose of the LPAD pathway.[133] An expert who attended our meeting later told us there is a great need to address how to develop narrow-spectrum antibiotics—those designed to treat a single or small number of bacterial pathogens—using LPAD. FDA held a public meeting in July 2019 to solicit stakeholder comments on the draft LPAD pathway guidance, and FDA officials told us

---

[130] Pub. L. No.114-255, § 3042, 130 Stat. 1112-14 (Dec. 13, 2016) (codified at 21 U.S.C. § 356). LPAD applies to eligible drugs and biological products regulated by Center for Drug Evaluation and Research and the Center for Biologics Evaluation and Research, respectively.

[131] Department of Health and Human Services, Food and Drug Administration, *Limited Population Pathway for Antibacterial and Antifungal Drugs (Draft Guidance for Industry)* (Silver Spring, Md.: June 2018).

[132] Drug developers seeking approval under the LPAD pathway may also seek QIDP designation and approval under other applicable provisions, such as accelerated approval, breakthrough therapy, or priority review.

As of August 2019, FDA had approved two drugs under the LPAD pathway: Arikayce for the treatment of lung disease caused by a group of bacteria in a limited population of patients who do not respond to conventional treatment, and Pretomanid for the treatment of a specific type of highly treatment-resistant tuberculosis of the lungs.

[133] In the draft LPAD guidance, FDA states its interpretation of "limited population" to mean "a group of patients that is limited in such a way that is clinically relevant to health care providers." Food and Drug Administration, *Limited Population Pathway for Antibacterial and Antifungal Drugs* (Draft Guidance for Industry): p.3.

they expect to finalize the guidance by February 2020. However, as of March 17, 2020, FDA had not yet issued final guidance.

In addition to issuing guidance, and to help inform future guidance, FDA engages with industry stakeholders to discuss and identify possible solutions to challenges related to the clinical development of antibiotics and nontraditional products. For example, FDA has held multiple public workshops, including one in November 2019 with experts from NIH's National Institute of Allergy and Infectious Diseases, the Infectious Disease Society of America, and The Pew Charitable Trusts to better understand the current state of antibiotic clinical trials in the United States, and how to enhance enrollment and research in these trials.

FDA officials told us they believe it is too early to issue guidance that would be broadly applicable and useful to nontraditional product developers. They explained that for certain types of nontraditional products, the approaches and specifics of product development are varied and evolving quickly. Instead, FDA's Center for Biologics Evaluation and Research has a program in place that allows developers to meet with FDA prior to beginning clinical trials to obtain advice on a wide range of development-related topics.[134]

Implemented Medicare Payment Mechanisms

CMS uses Medicare payment mechanisms to help increase reimbursement to hospitals for certain antibiotics. For qualifying antibiotics, these payments are a form of indirect pull incentive because they have the potential to increase the demand for the new antibiotics after they reach the market, which could in turn improve their financial performance. Beginning in fiscal year 2020, CMS updated how it will pay hospitals for treating Medicare patients who have an antibiotic-resistant infection.[135] Specifically, CMS changed the eligibility criteria and payment

---

[134] This program is called INTERACT—INitial Targeted Engagement for Regulatory Advice on CBER producTs.

[135] Department of Health and Human Services, Centers for Medicare & Medicaid Services, *Medicare Program; Hospital Inpatient Prospective Payment Systems for Acute Care Hospitals and the Long-Term Care Hospital Prospective Payment System and Policy Changes and Fiscal Year 2020 Rates; Quality Reporting Requirements for Specific*

amount for antibiotics that qualify for "new technology add-on payments" and how it pays hospitals for treating Medicare patients with antibiotic-resistant infections. These payment changes are:

- *Revised eligibility criteria for and amount of add-on payments.* New technology add-on payments provide hospitals with additional compensation for a period of 2 or 3 years when they use qualifying new technologies or drugs that offer substantially improved clinical treatment, and when regular Medicare payments for the hospital stay are inadequate to cover the cost of the new technology or drug.[136] Generally, medical services and technologies must be new and must demonstrate a substantial clinical improvement over existing services or technologies to receive the additional payment.[137] However, CMS has acknowledged the difficulty antibiotic developers face in demonstrating such substantial clinical improvement due to manufacturers seeking FDA approval for most antibiotics on the basis of noninferiority clinical trials, as described above. To make it easier for antibiotics to qualify for the additional payments, under the revisions to the CMS payment policy beginning in fiscal year 2021, CMS will consider all antibiotics with a QIDP designation from FDA to be "new" for purposes of the add-on payment, and these antibiotics will not have to meet the substantial clinical improvement criteria.[138]
- In addition, CMS has increased the amount of the temporary add-on payment for qualifying antibiotics. Prior to this change, the add-

---

Providers; Medicare and Medicaid Promoting Interoperability Programs Requirements for Eligible Hospitals and Critical Access Hospitals, 84 Fed. Reg. 42044 (Aug. 16, 2019).

136 The new technology add-on payments are made on top of the payment made under the inpatient prospective payment system. Under the inpatient prospective payment system, Medicare pays hospitals a single bundled payment per patient stay, which is based on multiple factors. 42 C.F.R. § 412.87 (2018).

137 In addition, the diagnosis related group rate otherwise applicable to discharges involving the medical service or technology must also be determined inadequate. 42 C.F.R. § 412.87(b) (2018).

138 84 Fed. Reg. 42044, 42292, and 42611 (adding a new paragraph (c) to 42 C.F.R. § 412.87) (Aug. 16, 2019).

on payments for qualifying antibiotics were limited to 50 percent of the cost of the drug. Under the new policy, the payment percentage increased to a maximum of 75 percent of the cost of the drug.[139] CMS has specified that two antibiotics are eligible for new technology add- on payments in fiscal year 2020.[140]

- *Increased payment for hospital stays.* CMS changed the severity level designation for certain antibiotic resistance-related diagnosis codes, in recognition of the added clinical complexity and cost of treating patients with antibiotic resistance.[141] This change in severity level can result in higher payments to hospitals when treating patients diagnosed with antibiotic resistance, which, according to the Administrator of CMS in an August 2019 blog post, will create "financial flexibility for physicians to prescribe the appropriate new antibiotics."[142] The Administrator also noted that CMS made this policy change because it recognized that new technology add-on payments are temporary and "further action was needed to realign financial incentives for antibiotics for the long-term."

See appendix III for additional examples of efforts to support antibiotic R&D by these and other federal agencies.

---

[139] 84 Fed. Reg. 42044, 42297, and 42612 (revising paragraphs (a) and (b) of 42 C.F.R. § 412.88 (Aug. 16, 2019).

[140] The antibiotics that qualify for Medicare new technology add-on payments in fiscal year 2020 are Vabomere and Zemdri. 84 Fed. Reg. 42044, 42188 (Vabomere), and 42191 (Zemdri) (Aug. 16, 2019).

[141] 84 Fed. Reg. 42044, 42150 (Aug. 16, 2019). CMS changed the severity designation for these diagnosis codes from "non-CC" to "CC," which recognizes the presence of a complication or comorbidity that requires the hospital to dedicate more resources for the care of that patient than typically needed for the specific diagnosis.

[142] S. Verma, "Aligning Payment and Prevention to Drive Antibiotic Innovation for Medicare Beneficiaries," *Health Affairs Blog*, August 2, 2019, https://www.healthaffairs.org/do/10.1377/hblog20190802.505113/full/.

Federal Efforts Have Not Fully Incentivized Antibiotic Development, and HHS Lacks a Strategy to Develop New Incentives

Experts and antibiotic developers told us that the economic challenges have remained despite the available federal push and pull incentives for antibiotic R&D. Currently available premarket push incentives include grants and awards from NIH and BARDA that fund antibiotic R&D; currently available postmarket pull incentives include the additional market exclusivity available through QIDP designation and Medicare add- on payments for antibiotics. (See Figure 3.) Both of the antibiotic companies that declared bankruptcy in 2019 had received push incentives from BARDA and pull incentives through Medicare New Technology Add-on Payments and the QIDP 5-year extension of market exclusivity.[143]

While experts at our meeting and antibiotic developers told us that push incentives have been helpful, they also said push incentives alone are not sufficient to sustain antibiotic development.

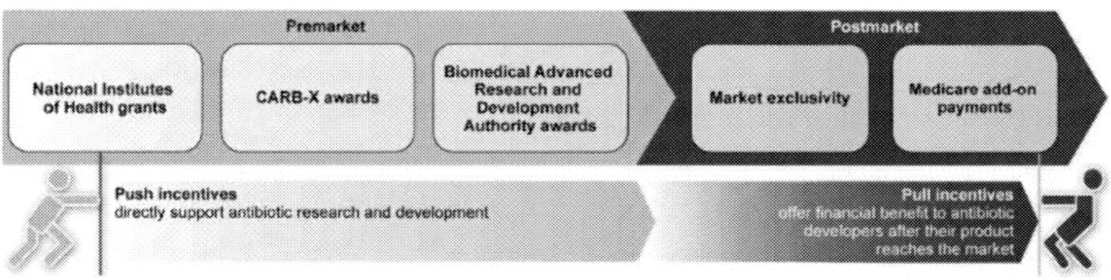

Source: GAO summary of information from the Department of Health and Human Services and literature. | GAO-20-341.

Figure 3. Examples of Currently Available Federal Incentives for Antibiotic Development.

[143] As of September 2019, one of the companies had received about $136 million and the other had received about $60 million from BARDA.

Data for Figure 3: Examples of currently available federal incentives for antibiotic development

Pre-market

- National Institutes of Health grants
- CARB-X awards
- Biomedical Advanced Research and Development Authority awards
- Push incentives directly support antibiotic research and development

Post market

- Market exclusivity
- Medicare add-on payments
- Pull incentives offer financial benefit to antibiotic developers after their product reaches the market

Note: CARB-X is the Combating Antibiotic-Resistant Bacteria Biopharmaceutical Accelerator.

For example, two antibiotic developers we spoke with explained that push incentives have provided needed funding for conducting R&D, but said that push incentives will not help cover the costs they will incur after their drug reaches the market— for example, to manufacture and market their product.[144] Experts and antibiotic developers have indicated that the effects of the existing pull incentives, QIDP market exclusivity, and Medicare add-on payments on stimulating development of new antibiotics have been limited for the following reasons:

- *QIDP and market exclusivity.* As we previously reported, several pharmaceutical companies told us that the market exclusivity incentive may not stimulate the development of new antibiotics, because the extension is unlikely to extend past the typical patent life of a new drug.[145] In addition, a representative from The Pew

---

[144] One expert at our meeting noted that antibiotic companies face a "second wave" of costs after a drug reaches the market—costs to manufacture the drug and to conduct additional clinical studies for populations and indications beyond the original FDA approval.

[145] Market exclusivity and patent life generally run concurrently. While market exclusivity can range from about 3 to 7 years from drug approval, according to FDA, a drug patent

Charitable Trusts said that, while the passage of the Generating Antibiotic Incentives Now Act initially bolstered private investments in antibiotics, it did not ultimately stabilize the pipeline of antibiotics in development, noting that since then, several large pharmaceutical companies have discontinued their antibiotics R&D programs.

- *Medicare updates to hospital payments.* While CMS recently increased new technology add-on payments for certain antibiotics beginning in fiscal year 2020 to help improve access to antibiotics, these payments are limited to antibiotics used to treat Medicare patients.[146] In addition, although Medicare increased the add-on payment amount to up to 75 percent of the estimated costs of qualifying antibiotics in excess of the regular Medicare payment, hospitals could still face costs for providing these drugs that are not covered by the Medicare payment. Furthermore, representatives from an antibiotic company and a biotechnology trade association told us the add-on payments do not directly incentivize hospital pharmacies to purchase the drug, because the add-on payment may not flow back to the pharmacy department's budget. For these reasons, it remains to be seen whether the Medicare new technology add-on payments to hospitals for inpatient antibiotics will help improve the return on investment for antibiotic developers and further stimulate the antibiotic

---

generally expires 20 years from the date the patent holder filed its application with the United States Patent and Trademark Office. See GAO, *Antibiotics: FDA Has Encouraged Development, but Needs to Clarify the Role of Draft Guidance and Develop Qualified Infectious Disease Product Guidance*, GAO-17-189 (Washington, D.C.: Jan. 31, 2017).

HHS stated in a 2017 report to Congress that it was still too early to assess whether the Generating Antibiotic Incentives Now Act was addressing the need for new antibiotics because all 12 of the approved QIDP drugs were already in development before the incentives were created. Department of Health and Human Services, *Generating Antibiotic Incentives Now: Required by Section 805 of the Food and Drug Administration Safety and Innovation Act. Public Law 112-144* (2017), p.12.

146 Department of Health and Human Services, Centers for Medicare & Medicaid Services, *Medicare Program; Hospital Inpatient Prospective Payment Systems for Acute Care Hospitals and the Long-Term Care Hospital Prospective Payment System and Policy Changes and Fiscal Year 2020 Rates; Quality Reporting Requirements for Specific Providers; Medicare and Medicaid Promoting Interoperability Programs Requirements for Eligible Hospitals and Critical Access Hospitals*, 84 Fed. Reg. 42044 (Aug. 16, 2019).

development pipeline. Similarly, it remains to be seen how CMS's policy change that provides increased payments for hospital stays when Medicare patients have been diagnosed with certain types of antibiotic-resistant infections will affect hospitals' use of new antibiotics.

In light of the limitation of existing incentives for antibiotic development, experts, federal officials, and antibiotics developers have called for additional postmarket pull incentives to reinvigorate the pipeline of antibiotics under development. For example, PACCARB issued recommendations to the Secretary of HHS in September 2017 and July 2019 for the adoption of pull incentives, calling for the development of market entry rewards and options for plausible business models.[147] In addition TATFAR—of which officials from BARDA, CDC, FDA, and NIH are members—reported that it is critical to develop a pull incentive strategy now to ensure that enough antibiotics are available in the future.[148] Former FDA Commissioner Dr. Scott Gottlieb also stated in 2018 that he was "deeply concerned that without stronger pull incentives that encourage more R&D, we'll see a far less robust pipeline of products than we need to address antimicrobial resistance."[149] Eight of the antibiotic developers we interviewed told us they think additional financial incentives are needed.[150]

---

[147] See section I-3.1 of Presidential Advisory Council on Combating Antibiotic-Resistant Bacteria, *Recommendations for Incentivizing the Development of Vaccines, Diagnostics, and Therapeutics*, and goal 4, priority 1 of *Priorities for the National Action Plan on Combating Antibiotic-Resistant Bacteria: 2020-2025: A Report with Recommendations* (July 2019).

[148] C. Ardal et al., "Pull Incentives for Antibacterial Drug Development: An Analysis by the Transatlantic Task Force on Antimicrobial Resistance," *Clinical Infectious Diseases*, vol. 65, no. 8 (2017). TATFAR was created in 2009 with the goal to improve international coordination of efforts to fight antibiotic resistance. It includes representatives from the United States, Canada, the European Union, and Norway. CDC currently serves as the secretariat for TATFAR.

[149] See S. Gottlieb, "FDA's Strategic Approach for Combating Antimicrobial Resistance" (speech), September 14, 2018, remarks as prepared for delivery, https://www.fda.gov/news-events/speeches-fda-officials/fdas-strategic-approach-combatin g-antimicrobial-resistance-09142018.

[150] We asked the developers, "what other incentives, if any, could be offered by U.S. federal agencies or other entities to promote the development of new antibiotics and nontraditional products?" Of the remaining three developers we interviewed, one did not respond to this

For example, one developer said that sales revenues from antibiotics will never be sufficient to justify R&D investments, and another noted that financial incentives are needed during the first few years after a new antibiotic reaches the market to cover not only these costs, but also to conduct additional clinical trials to help expand the drug's possible market. Finally, several experts at our expert meeting noted that, without pull incentives, most of the small companies currently developing antibiotics are unlikely to survive, and large pharmaceutical companies will likely continue to exit the antibiotic market.

Advisory groups and others have identified multiple options for how postmarket pull incentives could be designed, including market entry rewards—either in the form of lump sum payments or transferable vouchers that could be sold to confer additional market exclusivity to other pharmaceutical drugs—or reimbursement reform, such as licensing arrangements or add-on payments for hospital-administered antibiotics. (See Figure 4.) The four advisory groups whose papers we reviewed each recommended market entry rewards as effective pull incentive options.

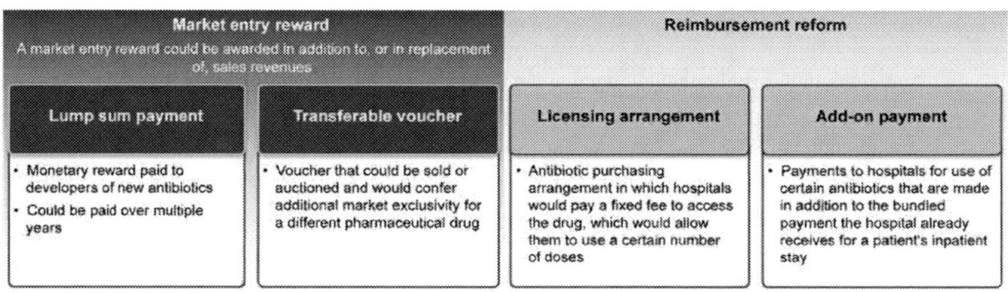

Source: GAO summary of publicly available products. | GAO-20-341.

Figure 4. Examples of Possible Postmarket Incentive Options to Encourage the Development of Antibiotics Identified by Advisory Groups and Others.

question, one discussed changes they would like FDA to make regarding clinical trials, and one said they could not think of any additional incentives.

Data for Figure 4: Examples of possible postmarket incentive options to encourage the development of antibiotics identified by advisory groups and others

Market entry reward

A market entry reward could be awarded in addition to, or in replacement of, sales revenues

Lump sum payment

- Monetary reward paid to developers of new antibiotics
- Could be paid over multiple years

Transferable voucher

- Voucher that could be sold or auctioned and would confer additional market exclusivity for a different pharmaceutical drug

Reimbursement reform

Licensing arrangement

- Antibiotic purchasing arrangement in which hospitals would pay a fixed fee to access the drug, which would allow them to use a certain number of doses

Add-On payment

Payments to hospitals for use of certain antibiotics that are made in addition to the bundled payment the hospital already receives for a patient's inpatient stay

While Commissioner of the FDA, Dr. Scott Gottlieb proposed an antibiotics licensing arrangement, which he called a subscription model, in a 2018 speech.

Views on the utility of reimbursement reform as a pull incentive strategy are mixed. For example, representatives from The Pew Charitable Trusts stated their view that, while CMS's recent changes to Medicare payment for antibiotics will likely be helpful to some degree, no reimbursement policy on its own would be able to increase antibiotic sales revenues sufficiently to transform the business model for antibiotics. An antibiotic developer we spoke to also told us that reimbursement policies would not be sufficient to support their business model because of low sales volumes for new antibiotics. The developer explained that it can take

2 or 3 years of antibiotic sales to recoup their R&D costs and finance their ongoing business operations, and that while larger pharmaceutical companies can rely on other profitable drugs to offset those costs, they could not because they did not have other drugs on the market. However, a representative from a biotechnology trade association told us that increasing reimbursement could help alleviate some of the economic challenges faced by developers of antibiotics that are already on or about to reach the market while policy makers explore longer-term pull incentive strategies. TATFAR cautioned that simply increasing reimbursement for antibiotics could potentially limit patient access, particularly for patients without health insurance—including those in low-and middle-income countries—and it could incentivize only antibiotics for common types of infections with a large market potential, rather than for rare, yet dangerous, types of pathogens.

Advisory groups and others have evaluated potential market entry reward models, taking into consideration factors such as format, value, funding sources, and eligibility criteria. Some have proposed that receipt of a market entry reward should be delinked, fully or partially, from sales revenues—that is, the developer would have to forgo some or all sales revenue as a condition of receiving the reward. Proponents of delinkage believe that separating revenues from antibiotics sales volumes would discourage aggressive sales that could lead to overuse. An expert who attended our meeting later told us that policies to incentivize use of new antibiotics must be balanced with policies to monitor prescribing of new drugs to prevent inappropriate use. Generally, advisory groups stipulate that to maximize the public health benefit, only antibiotics that treat what are deemed to be high priority bacteria should be eligible for a reward.[151] Specific recommendations and conclusions included the following:

- TATFAR concluded in 2017 that a partially delinked market entry reward of approximately $500 million would be the least

[151] C. M. Morel and S. E. Edwards, "Encouraging Sustainable Use of Antibiotics: A Commentary on the DRIVE-AB Recommended Innovation Incentives," *The Journal of Law, Medicine & Ethics*, vol. 46, no. S1 (2018).

disruptive option but noted that additional assessment would be necessary to select the most appropriate model and determine governance and other design elements.[152]

- PACCARB expressed support for a delinked model, in which a company accepting a market entry reward would be required to forgo marketing activities and profits based on sales volume. In addition, they suggested the establishment of an antibiotic incentive fund supported by an antibiotic usage fee or the sale or auction of transferable exclusivity vouchers as plausible options for financing pull incentives.
- The Duke University Margolis Center for Health Policy recommended in 2017 a delinked, public-private market entry reward model. This model was comprised of publicly funded market entry rewards for qualifying antibiotics for the first 5 or 6 years, followed by privately funded "value-based" contracts between antibiotic developers and health care payors, in which the payor could agree, for example, to pay a predetermined amount for full access to the antibiotics for a given population.[153] The Duke-Margolis Center proposal did not specify a funding source, but it noted multiple options for consideration, including general government funds, antibiotic use taxes, or the sale of transferable exclusivity vouchers.
- The European DRIVE-AB project recommended in 2018 an internationally funded, partially delinked market entry reward valued at approximately $1 billion per antibiotic, paid over the course of 5 or more years.[154] Recipients of a market entry reward

---

[152] C. Ardal et al., "Pull Incentives for Antibacterial Drug Development." TATFAR also published a literature review in 2016 that summarized twelve reports by researchers and advisory groups who analyzed possible antibiotic pull incentive strategies. See K. Sciaretta et al., "Economic Incentives for Antibacterial Drug Development: Literature Review and Considerations from the Transatlantic Task Force on Antimicrobial Resistance," *Clinical Infectious Diseases*, vol. 63, no. 11 (2016).

[153] Duke Margolis Center, *Value-Based Strategies for Encouraging New Development of Antimicrobial Drugs*.

[154] The DRIVE-AB project, which is the short name for the "Driving Re-investment in R&D and Responsible Antibiotic Use," was funded by the Innovative Medicines Initiative, which

would be allowed to sell their drug on the private market, but they would agree to certain marketing restrictions to discourage inappropriate use.

HHS may need to request authority and appropriations to create and implement certain types of market entry rewards. For example, HHS does not currently have authority to offer transferable exclusivity vouchers to antibiotic developers, since that would require a change in statute. Advisory groups also noted that the various pull incentive approaches would require additional public or private expenditures and offered possible sources of funding. For example, in addition to general fund revenues, PACCARB suggested that pull incentives could be funded through antibiotic usage fees, the auctioning of transferable exclusivity vouchers, or by allowing developers of new antibiotics to earn a transferable exclusivity voucher. The Duke-Margolis Center suggestions included funding market entry rewards through a yearly per-member fee for all health insurance plans. Transferable exclusivity vouchers may not require an independent funding source, because the value of the reward is based on the sale of the voucher to another drug developer. However, vouchers would still increase public and private health care expenditures, because expenditures would likely increase for drugs for which the extra period of exclusivity was purchased due to the delayed entry of lower-priced generics. Finally, reimbursement reform could increase health care expenditures for health care payors, including Medicare and private health insurance carriers.

Although PACCARB, TATFAR, and other experts have called for additional postmarket pull incentives to increase the antibiotic pipeline, as of January 2020 HHS has not developed a strategy for creating these incentives. HHS officials told us that the department created an interagency workgroup within HHS in spring 2019 to identify possible pull

received financial contributions from the European Union and the European Federation of Pharmaceutical Industries and Associations.

incentive options, among other things.[155] The recently convened HHS interagency workgroup is a step in the right direction toward exploring options for new antibiotic development incentives. Through this workgroup, HHS has an opportunity to determine which types of postmarket incentives it believes would most effectively incentivize the development of new treatments for antibiotic-resistant infections. However, it is unclear whether the HHS interagency workgroup's efforts will include consideration of such incentives because, according to HHS officials in January 2020, the interagency workgroup was still considering possible recommendations for HHS leadership and had not produced any specific documents to share with us.

The Government Performance and Results Act of 1993 (GPRA) and the GPRA Modernization Act of 2010, which significantly enhanced agencies' responsibilities under GPRA, include principles for federal agencies to consider related to developing strategies for achieving results, among other principles.[156] We have previously reported that these principles can serve as leading practices for planning at lower levels within agencies, such as individual programs or initiatives.[157] Our past work has shown that strategic frameworks can serve as a basis for guiding policy makers, including congressional decision makers and agency officials, when making decisions about resources, programs and activities, particularly in relation to issues that are national in scope, such as antibiotic development.[158] Developing a strategic framework that outlines new postmarket pull incentives and their key design elements—such as monetary value, eligibility criteria, and guidelines to prevent overuse—would be a first step toward identifying potential authorities and resources that may be needed to create the incentives, and toward determining

---

[155] In addition, to further assess the market economics for antibiotics, the HHS Office of the Assistant Secretary for Planning and Evaluation funded a study in 2018 to assess whether there are market failures that lead to suboptimal investment in antimicrobial drugs. The study is due to be completed in 2021.

[156] Pub. L. No. 103-62, 107 Stat. 285 and Pub. L. No. 111-352, 124 Stat. 3866.

[157] GAO-16-693.

[158] GAO, *Prescription Drugs: Strategic Framework Would Promote Accountability and Enhance Efforts to Enforce the Prohibitions on Personal Importation*, GAO-05-372 (Washington, D.C.: Sept. 8, 2005).

agency roles for implementation and oversight of the incentives. Until such incentives are developed, more drug companies may exit the antibiotic development sector, and the pipeline of new treatments for antibiotic-resistant infections may continue to decrease. Furthermore, the current significant federal investment in push incentives to support antibiotic R&D will remain a high-risk enterprise, if companies receiving large R&D grants are unable to sustain their business once their treatment reaches the market.

FEDERAL AGENCIES HAVE UNDERTAKEN SEVERAL EFFORTS TO PROMOTE THE APPROPRIATE USE OF ANTIBIOTICS, BUT KEY CHALLENGES REMAIN

Federal agencies have undertaken several efforts to promote the appropriate use of antibiotics through stewardship programs and activities. However, four key challenges remain that have limited this progress.

Federal Agencies Have Undertaken Several Efforts to Promote the Appropriate Use of Antibiotics through Stewardship Programs

To promote the appropriate use of antibiotics across health care settings through antibiotic stewardship programs and activities, federal agencies have undertaken several efforts that aim to reduce inappropriate antibiotic use, reduce health care costs, improve patient outcomes, and combat antibiotic resistance. Selected examples of these efforts are discussed below. (For more detailed information on agencies' efforts to promote the appropriate use of antibiotics, see app. IV.)

Published Requirements for Hospitals, Long-Term Care, and DOD and VA Facilities to Implement Antibiotic Stewardship Programs

Federal agencies require certain types of health care facilities to implement antibiotic stewardship programs, as follows:

- *CMS.* In September 2019, CMS finalized new health and safety requirements for hospitals and critical access hospitals to implement antibiotic stewardship programs by March 30, 2020, as a condition of their participation in the Medicare and Medicaid programs.[159] Under these requirements, hospitals and critical access hospitals are required, among other things, to implement these programs facility- wide (which includes emergency departments) and to adhere to nationally recognized antibiotic prescribing guidelines.[160] Nearly 3 years prior, CMS published similar requirements for nursing homes and skilled nursing

---

159 Department of Health and Human Services, Centers for Medicare & Medicaid Services, *Medicare and Medicaid Programs; Regulatory Provisions to Promote Program Efficiency, Transparency, and Burden Reduction; Fire Safety Requirements for Certain Dialysis Facilities; Hospital and Critical Access Hospital (CAH) Changes to Promote Innovation, Flexibility, and Improvement in Patient Care*, 84 Fed. Reg. 51732 (Sept. 30, 2019) (pertinent provisions to be codified at 42 C.F.R. §§ 482.42 (d), 485.640).
The final health and safety (also known as Conditions of Participation) requirements for antibiotic stewardship programs in hospitals also apply to inpatient rehabilitation facilities (which are considered post-acute care facilities) and long-term acute care hospitals. 84 Fed. Reg. 51732, 51780 (Sept. 30, 2019).
"Critical access hospital" is a designation given to eligible rural hospitals by states and certified by CMS. 42 U.S.C. § 1395i-4(c)(2)(B), 42 C.F.R. § 485.606(b). Congress created the critical access hospital designation through the Balanced Budget Act of 1997 in response to a string of rural hospital closures in the 1980s and early 1990s. Among other criteria, eligible hospitals must have 25 or fewer acute-care inpatient beds and be located more than 35 miles from another hospital, although exceptions may apply. *See* Balanced Budget Act of 1997, Pub. L. No. 105-33, § 4201, 111 Stat. 251, 369-71 (codified in pertinent part at 42 U.S.C. §§ 1395f(l), 1395i-4, 1395m(g), (l)(8)). As of October 2019, there were 1,349 critical access hospitals in the United States.

160 Professional medical societies, such as the American Academy of Pediatrics, publish nationally recognized antibiotic prescribing guidelines for their specialties. In addition, the Infectious Diseases Society of America and the Society for Healthcare Epidemiology of America have jointly published evidence-based guidelines for implementing an antibiotic stewardship program in acute inpatient, long-term care, and emergency department care settings. See, for example, T. F. Barlam et al., "Implementing an Antibiotic Stewardship Program: Guidelines by the Infectious Diseases Society of America and the Society for Healthcare Epidemiology of America," *Clinical Infectious Diseases,* vol. 62, no. 10 (2016): pp. e51–e77.

facilities—collectively known as long-term care facilities—to establish antibiotic stewardship programs by December 4, 2017.[161] Experts, including those at our meeting and the PACCARB, credit these requirements with being a powerful lever for promoting the appropriate use of antibiotics; Medicare comprises a significant portion of the nation's health care expenditures—$741 billion in 2018, covering 59.9 million beneficiaries.[162]

- *DOD.* DOD published a policy, effective October 2017, requiring the establishment of antibiotic stewardship programs within its military medical treatment facilities and, one year later, issued guidance for implementation.[163] Among other things, the policy specified that these facilities' antibiotic stewardship programs include components such as (1) leadership commitment by each facility; (2) accountability; (3) pharmacy expertise, including antibiotic prescribing and use evaluation; (4) implementation of action for change that would demonstrate commitment to the program; and (5) training for clinicians regarding antibiotic resistance and prescribing practices. DOD officials told us that all of these facilities (both inpatient and outpatient) were in different stages of implementing the antibiotic stewardship policy.
- *VA.* In January 2019, VA updated its 2014 policy directive for the implementation and maintenance of antibiotic stewardship programs in its health care facilities, which provide both inpatient

---

161 Department of Health and Human Services, Centers for Medicare & Medicaid Services, *Medicare and Medicaid Programs; Reform of Requirements for Long-Term Care Facilities*, 81 Fed. Reg. 68688, 68697 (Oct. 4, 2016) (pertinent provision codified at 42 C.F.R. § 483.80(a)(3) (2018)).

162 Centers for Medicare & Medicaid Services, "Trustees Report & Trust Funds," accessed September 15, 2019, https://www.cms.gov/Research-Statistics-Data-and-Systems/Statistics-Trends-and-Report s/ReportsTrustFunds/index.

163 Department of Defense, DOD Instruction 6025.26, "DOD Program for Combating Antibiotic-Resistant Bacteria (CARB)" (Oct. 31, 2017) and Defense Health Agency Procedural Instruction Number 6025.09, "DOD Program for Combating Antibiotic- Resistant Bacteria (CARB)" (Oct. 2, 2018).
DOD's military medical treatment facilities include hospitals and medical centers located at military installations across the United States and abroad, plus ambulatory care clinics and dental clinics, and provide health care for active-duty service members, their dependents, and other eligible beneficiaries.

and outpatient services to veterans.[164] This policy directive includes requirements for its facilities to develop a written policy, conduct an annual evaluation of stewardship activities, ensure that adequate staff and resources are in place, and identify medical and pharmacy personnel as stewardship "champions." According to department officials, VA has successfully implemented antibiotic stewardship programs in all of its health care facilities.

Developed Incentives for Clinicians to Implement Antibiotic Stewardship Activities

CMS has developed incentives for eligible clinicians in any type of health care facility to improve antibiotic use and stewardship, as part of the agency's broader efforts to improve care for Medicare patients.[165] Through the Merit-based Incentive Payment System (MIPS) launched in 2017, CMS offers hundreds of quality measures and nearly 100 "improvement activities" on a wide range of topics—including the appropriate use of antibiotics—on which eligible clinicians can choose to report their performance to the agency.[166] CMS then adjusts payments higher for clinicians who report data and achieve a performance-based, final score above a certain threshold—and penalizes clinicians who do not achieve that threshold with lower payments.

---

[164] Department of Veterans Affairs, Veterans Health Administration Directive 1031, "Antimicrobial Stewardship Programs (ASP)" (Washington, D.C.: Jan. 30, 2019).

[165] The Medicare Access and CHIP Reauthorization Act of 2015 required CMS to implement an incentive program for clinicians participating in Medicare to receive higher payments based on their performance. The Medicare Access and CHIP Reauthorization Act of 2015, Pub. L. No. 114-10, 129 Stat. 87, 92 (Apr. 16, 2015), codified at 42 U.S.C. § 1395w-4.

[166] MIPS-eligible clinician types include physicians (doctors of medicine, which includes many specialties; doctors of dental surgery or dental medicine; and doctors of osteopathy), physician assistants, nurse practitioners, and others. In addition, to be eligible for MIPS in 2019, the clinician must (1) bill $90,000 or more in allowed charges for professional services covered under the Medicare Physician Fee Schedule; (2) provide more than 200 covered professional services; and (3) furnish covered professional services to more than 200 Medicare beneficiaries. Eligible clinicians may participate in MIPS as individuals or as part of a group that includes one or more of the eligible clinician types.
For more information on MIPS, see GAO, *Health Care Quality: HHS Should Set Priorities and Comprehensively Plan Its Efforts to Better Align Health Quality Measures*, GAO-17-5 (Washington, D.C.: Oct. 13, 2016) and *Medicare: Small and Rural Practices' Experiences in Previous Programs and Expected Performance in the Merit-based Incentive Payment System*, GAO-18-428 (Washington, D.C.: May 31, 2018).

Published Guidance on Implementing Antibiotic Stewardship Programs

Federal agencies have published guidance for health care facilities on how to implement antibiotic stewardship, as follows:

- *AHRQ.* Through a 5-year nationwide project, the AHRQ Safety Program for Improving Antibiotic Use has provided technical assistance and CDC's guidance to hospitals, long-term care settings, and physicians' offices to promote implementation of antibiotic stewardship activities and help clinicians select optimal antibiotic treatment regimens. In December 2018, AHRQ completed implementation of this guidance in more than 400 hospitals, which included six DOD facilities and 79 critical access hospitals, according to AHRQ officials.
- **CDC.** Since 2014, CDC has published a series of guidance documents—called the Core Elements of Antibiotic Stewardship (Core Elements)—to promote the appropriate use of antibiotics in health care.[167] The Core Elements are tailored to hospitals, nursing homes, outpatient settings, small and critical access hospitals, and low- and middle-income countries with limited resources. Common elements in these guidance documents include (1) leadership commitment, (2) implementation of policies and interventions to improve antibiotic use, (3) tracking and reporting antibiotic use, and (4) education to providers on appropriate antibiotic use.

---

[167] Department of Health and Human Services, Centers for Disease Control and Prevention, *The Core Elements of Hospital Antibiotic Stewardship Programs* (Atlanta, Ga.: 2014) and *The Core Elements of Hospital Antibiotic Stewardship Programs: 2019* (Atlanta, Ga.: 2019), *The Core Elements of Antibiotic Stewardship for Nursing Homes* (Atlanta, Ga.: 2015), *The Core Elements of Outpatient Antibiotic Stewardship* (Atlanta, Ga.: 2016), *Implementation of Antibiotic Stewardship Core Elements at Small and Critical Access Hospitals* (Atlanta, Ga.: 2017), and *The Core Elements of Human Antibiotic Stewardship Programs in Resource-Limited Settings: National and Hospital Levels* (Atlanta, Ga.: 2018).

Expanded the Collection of Antibiotic Use Data

Since we reported on antibiotic use data gaps in 2011, CDC has expanded its collection of such data regarding both inpatient and outpatient settings through its own surveillance systems, as well as from other sources.[168] In particular, CDC has focused its efforts to expand antibiotic use data collection from hospitals, where an estimated one in two patients receives an antibiotic for at least one day during an average hospital stay.[169] CDC launched its AU Option in 2011 as a voluntary, electronic reporting tool added on to the pre-existing NHSN.[170] The AU Option allows the nation's 6,849 hospitals that are already reporting to the NHSN to submit their antibiotic use data in a standardized format.[171] CDC then aggregates such data to calculate national benchmarks and allows hospitals to compare their actual antibiotic use against those benchmarks. In addition, CDC has periodically conducted prevalence surveys through the EIP to gather data on health care-associated infections and antibiotic use in about 200 hospitals and 161 nursing homes in 10 states. With regard to outpatient settings, CDC has acquired, through a proprietary source, 8 years of pharmacy data on antibiotic prescriptions since 2011, which the agency is using to better characterize patterns in outpatient prescribing and to develop targeted interventions for high-prescribing areas.

---

[168] GAO, *Antibiotic Resistance: Data Gaps Will Remain Despite HHS Taking Steps*, GAO-11-406 (Washington, D.C.: June 1, 2011).

[169] J. Baggs et al., "Estimating National Trends in Inpatient Antibiotic Use among U.S. Hospitals from 2006 to 2012," *JAMA Internal Medicine*, vol. 176, no. 11 (2016): pp. 1639-1648.

[170] CDC data indicate that, as of January 1, 2020, there were 6,849 hospitals, including 13 DOD hospitals located on military bases outside the United States, that were eligible for reporting data to the AU Option. Although participation in the NHSN AU Option is generally voluntary, DOD and VA require their hospitals to report antibiotic use data to the AU Option. DOD officials told us that 44 DOD hospitals were reporting such data, as of September 30, 2019, and VA officials told us that 113 VA hospitals were reporting such data, as of January 1, 2020.

[171] As previously noted, hospital types that report data to NHSN include general acute care hospitals, critical access hospitals, children's hospitals, oncology hospitals, long-term acute care hospitals, and inpatient rehabilitation facilities. The AU Option is part of a reporting module called the AUR Module that also tracks antibiotic resistance through the AR Option.

Developed Antibiotic Stewardship Training for Various Health Care Settings

Federal agencies have developed training on antibiotic stewardship, as follows:

- *CDC.* In 2018, CDC launched a free, online training course for various types of clinicians—including physicians, dentists, pharmacists, physician assistants, and nurses—to inform them about proper antibiotic prescribing and strategies for communicating with patients.[172] Clinicians can receive credit for partial completion (at least 50 percent) or full completion of this training as improvement activities under MIPS in 2019.
- *CMS.* CMS has provided training, technical assistance, and other learning opportunities to more than 4,000 hospitals, 2,400 nursing homes, and 7,600 outpatient settings on best practices for antibiotic stewardship and guidance on *C. difficile* prevention. In addition, CMS and CDC have developed and launched free, online training to help nursing homes implement antibiotic stewardship and prevent and manage *C. difficile* infections.
- *DOD and VA.* These departments have also offered antibiotic stewardship training to their health care facilities through webinars, workshops, or briefings.

Funded Research

Federal agencies have funded research on antibiotic stewardship, as follows:

- *AHRQ.* Since 2015, AHRQ has increased its support for research to develop improved methods to combat antibiotic resistance and promote antibiotic stewardship, including through grants for research that will total more than $57 million, according to AHRQ

---

[172] As of late July 2019, nearly 16,000 individuals had registered for this training, according to CDC officials.

officials. This research includes studies on the role of diagnostic tools in improving antibiotic use and reducing antibiotic resistance. AHRQ has also published numerous research studies on antibiotic or antimicrobial stewardship that the agency funded or authored.

- *CDC.* CDC supports research to identify, develop, and implement practices to stop the spread of resistance and to promote appropriate use of antibiotics in health care. CDC also supports research to fill gaps in knowledge related to aspects of antibiotic use and resistance that have public health impact. According to agency officials, CDC has provided approximately $110 million since 2016 to support this research through cooperative agreements and contracts.

Continued National Public Awareness Campaign

In 2017, CDC revised a national campaign to promote public awareness about appropriate antibiotic use. The campaign, called "Be Antibiotics Aware: Smart Use, Best Care," is aimed at both health care providers and the general public and refines the message from CDC's earlier campaign ("Get Smart: Know When Antibiotics Work").

Collaborated Internationally

HHS's Office of Global Affairs has collaborated with other countries, including those participating in the TATFAR program, to promote the appropriate use of antibiotics internationally. In addition, CDC and the Office of Global Affairs launched the Antimicrobial Resistance Challenge at the United Nations General Assembly in September 2018 to catalyze global action against antibiotic resistance. A year later, CDC announced this challenge had resulted in nearly 350 commitments from government health officials, pharmaceutical and health insurance companies, and others from 33 countries to make formal commitments that further the progress against antimicrobial resistance, such as by improving appropriate antibiotic use.

Four Key Challenges Have Limited Federal Efforts to Promote the Appropriate Use of Antibiotics

We identified four key challenges that have limited progress in federal efforts to promote the appropriate use of antibiotics, based on our analysis of documents, interviews with agency officials and experts, and other information. First, federal requirements for antibiotic stewardship programs apply only to certain types of health care facilities, and federal incentives for clinicians to adopt antibiotic stewardship activities are optional, limiting implementation of antibiotic stewardship across the health care spectrum. Second, CDC faces challenges in collecting complete antibiotic use data, limiting the agency's ability to monitor and improve antibiotic use. Third, the CARB Task Force has not identified and reported on agencies' plans to address the challenges related to expanding antibiotic stewardship programs and antibiotic use data collection across health care settings, so these plans are not publicly known. Fourth, antibiotic stewardship training for health care providers may have limited success in improving antibiotic prescribing behavior, and federal agencies indicate that it is challenging to evaluate the effectiveness of such training.

Federal Requirements and Incentives Are Limited

Federal requirements for antibiotic stewardship programs are limited to certain types of health care facilities, and federal incentives for antibiotic stewardship activities are optional and limited to eligible Medicare clinicians, such as physicians.

- *Federal requirements for antibiotic stewardship programs are limited to certain types of health care facilities.* As previously noted, federal requirements for antibiotic stewardship programs are currently limited to hospitals and critical access hospitals, long-term care facilities such as nursing homes, and DOD and VA health care facilities. However, CMS has not yet developed requirements for ambulatory surgery centers or dialysis centers to implement antibiotic stewardship programs, which the National

Action Plan called for being implemented by March 2018. CMS officials told us that the agency would develop those requirements once the rule for hospitals and critical access hospitals—which was delayed—was finalized. In addition, CMS's health and safety requirements do not extend to other types of outpatient settings—such as physicians' offices, retail clinics, and urgent care centers[173]—where inappropriate antibiotic use has been found to be high.[174]

In the absence of regulatory levers, CDC and AHRQ encourage those types of facilities to establish antibiotic stewardship programs on a voluntary basis. Experts, including those at our meeting, indicate that expansion of antibiotic stewardship across the health care spectrum is likely to remain limited without additional federal requirements or other meaningful incentives—thus hindering the nation from fully achieving the benefits of appropriate antibiotic use. Such benefits include better patient outcomes, lower health care costs, and slower growth of antibiotic resistance.

- *CMS incentives for clinicians to improve antibiotic use are optional, and implementation has been limited.* The MIPS program's effect on incentivizing appropriate use of antibiotics is limited, in part, because the incentives are available only to clinicians who meet MIPS eligibility criteria and because eligible

---

[173] CMS conditions payments for Medicare services to inpatient facilities, such as hospitals and long-term care facilities, and certain outpatient facilities, such as ambulatory surgical centers and dialysis centers, on compliance with CMS's health and safety standards (i.e., CMS's Conditions of Participation or Conditions for Coverage requirements). In addition, CMS generally pays clinicians separately for their Medicare services, regardless of whether the treatment takes place in health care facilities for which CMS has established health and safety standards, or in facilities that are not subject to CMS's health and safety standards, such as physician offices, retail clinics, and urgent care centers.

[174] As previously noted, CDC has reported that at least 30 percent, and as much as 50 percent, of antibiotic use in outpatient settings is inappropriate in the United States. See also D. L. Palms et al., "Comparison of Antibiotic Prescribing in Retail Clinics, Urgent Care Centers, Emergency Departments, and Traditional Ambulatory Care Settings in the United States," *JAMA Internal Medicine*, vol. 178, no. 9 (2018): pp. 1267-1269; and D. L. Palms et al., "First-Line Antibiotic Selection in Outpatient Settings," *Antimicrobial Agents and Chemotherapy*, vol. 63, no. 11 (2019): pp. e01060-19.

clinicians can choose not to report data to CMS. In addition, participating clinicians have a wide range and number of quality measures and improvement activities, beyond those related to antibiotics, from which the clinicians can choose to report data to CMS to meet program requirements; thus, the likelihood that clinicians will choose to report on antibiotics- related measures or activities may remain low. For example, in 2017, MIPS-eligible clinicians were generally required to select and submit data to CMS on six out of 271 available quality measures; we identified nine of those measures as being related to antibiotics. MIPS-eligible clinicians were also generally required to select and submit data that year for up to four out of 93 available improvement activities; we identified one such activity as being related to antibiotics.

Our analysis of CMS data on MIPS participation in 2017, the program's first performance year and the most recently available data, indicates that implementation of the antibiotics-related quality measures and improvement activities was limited. According to a CMS report, a total of 1,057,824 clinicians were eligible for MIPS in 2017, of which 1,006,319 clinicians, or 95 percent, reported data.[175] Based on our analysis of data contained in the CMS report's appendix, the number of 2017 MIPS-participating clinicians who reported to CMS on the nine antibiotics-related quality measures ranged from 844 clinicians to 33,631 clinicians; the measure on appropriate treatment for children with an upper respiratory infection was the most reported antibiotics-related measure.[176] By contrast, the most frequently

---

[175] Department of Health and Human Services, Centers for Medicare & Medicaid Services, *2017 Quality Payment Program Reporting Experience* (Baltimore, Md.: March 2019). CMS officials told us that for the first year of MIPS (the 2017 performance period), the agency estimates there were approximately 538,000 clinicians who were not eligible to participate in MIPS.

[176] We recognize that not all types of MIPS-eligible clinicians prescribe antibiotics or treat patient populations that relate to specific MIPS antibiotics-related quality measures; thus, such measures would only be appropriate for some providers to select and report on. For example, pediatricians and other clinicians who treat children would be more likely to

reported quality measures overall in 2017 were controlling high blood pressure (510,723 clinicians), preventive care and screening for tobacco use (492,357), and breast cancer screening (473,819). CMS's data also show that for the 2017 MIPS improvement activities, 47,645 of the 1,006,319 participating clinicians reported on the one improvement activity related to antibiotics that year: implementation of an antibiotic stewardship program.[177] Specifically, this activity referred to implementation of an antibiotic stewardship program that measured the appropriate use of antibiotics for several different conditions (upper respiratory infections in children, pharyngitis, and bronchitis in adults), according to clinical guidelines for diagnostics and therapeutics.[178]

CDC Faces Challenges in Collecting Complete Antibiotic Use Data, Limiting the Agency's Ability to Monitor and Improve Appropriate Use

CDC's ability to monitor and improve appropriate antibiotic use is limited by challenges it faces in collecting complete antibiotic use data across health care settings. According to CDC, experts we interviewed, and documents we reviewed, more data are needed to identify the extent of antibiotic use, including inappropriate use. In turn, CDC and experts say that more antibiotic use data would enable health care providers, federal agencies, and others to identify and target areas for improvement, track

---

choose and report on measures related to the appropriate use of antibiotics in children. In addition, it is possible that some MIPS-eligible clinicians implemented antibiotics-related quality measures but chose not to report them to CMS. However, the extent of these circumstances is unknown because CMS's data did not capture this information.

[177] We recognize that not all types of MIPS-eligible clinicians prescribe antibiotics; thus, MIPS antibiotics-related improvement activities would only be appropriate for some providers to select and report on. For example, clinicians who do not prescribe antibiotics would be unable to choose and report on an improvement activity to implement antibiotic stewardship programs. In addition, it is possible that some MIPS-eligible providers implemented antibiotic stewardship programs but chose not to report that to CMS as a MIPS improvement activity. However, the extent of these circumstances is unknown because CMS's data did not capture this information.

[178] In contrast, the most frequently reported improvement activities overall in 2017 were providing 24/7 access to eligible clinicians or groups that have real-time access to a patient's medical record (190,510 clinicians), use of decision support and standardized treatment protocols (118,450), and patient-centered medical home attestation (110,057).

results over time, and adjust antibiotic stewardship activities as needed.[179] We have also previously reported that monitoring antibiotic use over time in both inpatient and outpatient settings is important for understanding patterns in antibiotic resistance and for targeting stewardship activities.[180] In addition, WHO notes that data on global antibiotic use is essential for obtaining a comprehensive picture of antibiotic resistance and for identifying areas where actions are needed.[181]

Despite progress in collecting antibiotic use data (as previously discussed), CDC faces several challenges in its efforts to collect complete antibiotic use data. For example, health care providers across various inpatient and outpatient settings do not record such data in one centralized, electronic database. In addition, CDC officials told us that there are no uniform requirements at the federal level (with the exception of DOD and VA hospitals) for providers to report their antibiotic use data to a centralized database such as the NHSN AU Option, and, according to CDC officials and experts we interviewed, data collection can be costly for CDC and health care providers. Because of these and other challenges, CDC relies on data voluntarily reported by hospitals through the AU Option, and the agency collects its own data or purchases proprietary pharmacy data to estimate antibiotic use—and, to some degree, to assess appropriateness of use—across health care settings.

---

[179] For example, recent studies show that antibiotics are frequently inappropriately selected for common outpatient infections and that fluoroquinolones are frequently used inappropriately in adults, pointing to key areas where antibiotic stewardship could be targeted. See A. L. Hersh et al., "Frequency of First-line Antibiotic Selection among U.S. Ambulatory Care Visits for Otitis Media, Sinusitis, and Pharyngitis," *JAMA Internal Medicine,* vol. 176, no.12 (2016): pp. 1870-1872; and S. Kabbani et al., "Opportunities to Improve Fluoroquinolone Prescribing in the United States for Adult Ambulatory Care Visits," *Clinical Infectious Diseases*, vol. 67, no. 1 (2018): pp. 134-136.

[180] GAO-11-406.

[181] World Health Organization, *WHO Report on Surveillance of Antibiotic Consumption: 2016-2018 Early Implementation* (Geneva, Switzerland: 2018).

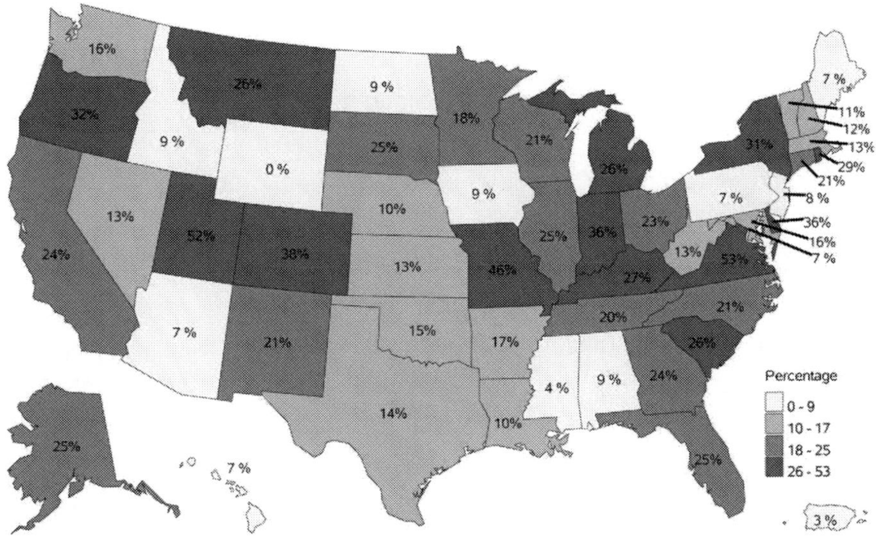

Source: Centers for Disease Control and Prevention (CDC). | GAO-20-341.

Note: U.S. hospitals can choose to report their antibiotic use data, on a voluntary basis, to CDC through the National Healthcare Safety Network's Antimicrobial Use Option. The percentages of hospitals shown in this figure had electronically submitted at least one month of antibiotic use data to this database as of August 2019, according to CDC.

Figure 5. Percentage of U.S. Hospitals Reporting Antibiotic Use Data to CDC's National Healthcare Safety Network, by State, plus the District of Columbia and Puerto Rico, as of August 2019.

However, these data are incomplete owing to several limitations, as described by type of setting below.

- *Hospitals.* Our analysis of CDC data shows that although the number of hospitals participating in the AU Option has gradually risen since its launch in 2011, participation remains limited, with 1,561, or 23 percent, of the 6,849 eligible hospitals reporting at least one month of antibiotic use data as of January 1, 2020.[182] (See Figure 5 for a map showing the percentage of U.S. hospitals

---

[182] CDC data indicate that, as of January 1, 2020, there were 6,849 hospitals, including 13 DOD hospitals located on military bases outside the United States, that were eligible for reporting data to the AU Option.

reporting antibiotic use data to the AU Option, by state, plus the District of Columbia and Puerto Rico, as of August 2019.) While CDC officials told us they considered this level of participation to be an accomplishment given that participation is voluntary, the National Action Plan set 95 percent participation in the AU Option by 2020 as a significant outcome to support the plan's goal to strengthen national surveillance efforts to combat resistance.[183]

Experts, including those at our meeting, cite multiple challenges that CDC faces in collecting hospitals' antibiotic use data through the AU Option. For example, The Pew Charitable Trusts has stated that current, voluntary data are limited and that mandatory reporting would provide the data needed to establish a more accurate baseline of antibiotic use, identify stewardship interventions that would be most effective, and measure progress toward reducing inappropriate prescribing.[184] An expert who attended our meeting later suggested that CMS could implement a pay-for-reporting program to incentivize hospitals to report data to the AU Option, and that the program could transition to a pay-for-performance program over time.[185] In addition, experts we interviewed told us that a participating hospital must be willing to spend as much as tens of thousands of dollars for a vendor to customize software for their electronic health record systems to use the AU Option, in addition to investing time training staff on how to use it. CDC officials also told us that the agency lacks the authority to require hospitals to report their antibiotic use data, and that there is currently no federal funding available to assist hospitals with the investment needed to participate in the AU Option. Furthermore, hospitals' voluntary participation in the AU Option may remain limited until CDC's

---

[183] As previously noted, the National Action Plan contains five goals that are supported by numerous objectives, sub-objectives, agency-specific milestones, and significant outcomes.

[184] The Pew Charitable Trusts, *Trends in Antibiotic Use, 2018* (Aug. 29, 2018).

[185] Since the early 2000s, CMS has offered financial incentives to Medicare providers who report their performance on specified quality measures. Under a pay-for-reporting program, such as the Hospital Inpatient Quality Reporting program, a hospital may receive higher payments for reporting data on the quality measures used in the program. Under a pay-for-performance program, a hospital may receive higher payments based on its level of performance on the measures.

benchmark measures are adequately risk- adjusted for different locations and patient populations.[186] For example, one expert we interviewed said that because the AU Option currently aggregates data on the volume of antibiotics used without adequate risk adjustment, a hospital with a patient population that might warrant higher use of antibiotics may be reluctant to report its antibiotic use data to avoid looking like an unnecessarily high prescriber.[187] Regarding another data source for antibiotic use in hospitals, CDC's EIP provides more granular data at the patient level that allows CDC to assess the appropriateness of antibiotic use. However, CDC officials told us that the agency has been unable to repeat its hospital prevalence survey since 2015 due to insufficient resources (the next survey is expected in 2020) and that the survey encompasses a limited number of hospitals, patients, and states.

- *Nursing homes.* According to CDC, nursing homes may be the most challenging health care setting from which the agency collects antibiotic use data; CDC officials stated that this is because electronic health record systems, from which data could

---

[186] In 2016, CMS invited public comment on the possibility of CDC's benchmark measures' future inclusion in CMS's Hospital Inpatient Quality Reporting program, which would allow the public to access antibiotic use information on individual hospitals. Department of Health and Human Services, Centers for Medicare & Medicaid Services, *Medicare Program; Hospital Inpatient Prospective Payment Systems for Acute Care Hospitals and the Long-Term Care Hospital Prospective Payment System and Proposed Policy Changes and Fiscal Year 2017 Rates; Quality Reporting Requirements for Specific Providers; Graduate Medical Education; Hospital Notification Procedures Applicable to Beneficiaries Receiving Observation Services; and Technical Changes Relating to Costs to Organizations and Medicare Cost Reports,* 81 Fed. Reg. 25197 (Apr. 27, 2016). However, CDC officials told us that it would likely be a few years before the benchmark measures would be adequately risk-adjusted and therefore ready to become a required reporting measure.

[187] The AU Option's benchmark measures allow a participating hospital to compare its "observed" antibiotic use with nationally aggregated ("predicted") use for specific antibacterial agents administered to adult and pediatric patients in specific ward and intensive care unit locations. CDC notes that the measures are designed to serve as high-value targets or high-level indicators for hospitals' antibiotic stewardship programs. Thus, a given hospital's results are intended to prompt analysis of possible overuse, underuse, or inappropriate use of antibiotics; identify opportunities for improvement; and gauge the impact of stewardship efforts. CDC also notes that higher-than-average antibiotic use might be justified, while lower-than-average use might harm patients, and that additional analyses to determine the appropriateness of antibiotic use in individual instances are likely to require access to detailed, patient-level data that is beyond the scope of data collection and analysis using the AU Option.

be easily accessed, are less common in nursing homes.[188] In addition, CDC officials stated that the agency's collection of antibiotic use data through the EIP nursing homes prevalence survey has been limited in scope and frequency due to insufficient resources.

- *Outpatient settings.* Collecting data for outpatient settings, such as retail pharmacies, is also challenging. For example, CDC officials stated that one proprietary source from which CDC purchases data reflects the volume of pharmacy antibiotic prescriptions, but the data do not contain diagnostic information, preventing the agency from evaluating the appropriateness of those prescriptions. Other CDC or proprietary data sources from which the agency collects or purchases antibiotic use data are limited by the frequency with which those sources release such data, the age range of patients included in the data (i.e., whether they are over or under 65 years), or other characteristics. As previously noted, approximately 85 to 95 percent of the nation's antibiotic use, by volume, occurred in outpatient settings from 2010 through 2015.

The CARB Task Force Has Not Identified Plans to Address Challenges Related to Expanding Stewardship Programs and Antibiotic Use Data Collection

The National Action Plan calls for strengthening antibiotic stewardship and for the timely reporting of antibiotic use data across health care settings. Executive Order No. 13676, as previously noted, directs the CARB Task Force to provide annual updates to the President on federal government actions to combat antibiotic resistance, including progress made in implementing the National Action Plan and plans for addressing any barriers preventing its full implementation.[189] These annual updates are

---

[188] However, CDC officials stated that the agency has partnered with electronic health record vendors and pharmacies to obtain access to and analyze nursing home antibiotic use data to inform stewardship efforts.

[189] Exec. Order No. 13676, § 3(c)(ii) (Sept. 23, 2014). The National Action Plan contains a sub-objective on strengthening antibiotic stewardship in inpatient, outpatient, and long- term care settings by expanding existing programs, developing new ones, and monitoring

to include specific goals, milestones, and metrics for proposed actions and recommendations, taking into consideration federal resources.

However, in its progress reports covering the first four years of the National Action Plan's implementation—which were provided to the President and the public—the CARB Task Force has not identified plans to address barriers that agencies face in expanding antibiotic stewardship programs across health care settings.[190] For example, the task force did not include in the progress reports CMS's plans to address barriers to expanding its requirements for antibiotic stewardship programs in hospitals, which were delayed, or in certain other types of health care facilities.[191]

In addition, in its progress reports to date, the CARB Task Force has not identified plans to address the barriers to expanding the collection of antibiotic use data across health care settings.[192] For example, the task force did not include in the progress reports CDC's plans to address barriers to achieving the significant outcome of 95 percent of eligible hospitals participating in the AU Option by 2020, although participation was only 23 percent as of January 1, 2020. The CARB Task Force

---

progress and efficacy. In addition, the National Action Plan contains a 3-year milestone, which CMS missed, for the agency to require hospitals and certain other health care settings (long-term acute care hospitals, other post-acute facilities, ambulatory surgery centers, and dialysis centers) to implement antibiotic stewardship programs through the agency's Conditions of Participation.

For the purposes of this report, we consider the terms challenges and barriers to be synonymous.

190 CARB Task Force, *National Action Plan: Progress Report for Years 1 and 2*, *National Action Plan: Progress Report: Year 3*, and *National Action Plan: Progress Report: Year 4*.

191 CMS's notice explained that the extension of the timeline for final rule publication was due to the complexity of the rule and its substantive nature. Department of Health and Human Services, Centers for Medicare & Medicaid Services, *Medicare and Medicaid Programs; Hospital and Critical Access Hospital (CAH) Changes To Promote Innovation, Flexibility, and Improvement in Patient Care; Extension of Timeline for Publication of the Final Rule*, 84 Fed. Reg. 27069 (June 11, 2019). See 42 U.S.C. § 1395hh(a)(3)(B).

192 One of the National Action Plan's objectives is to expand and strengthen the national infrastructure for public health surveillance and data reporting and to provide incentives for timely reporting of antibiotic resistance and antibiotic use in all health care settings. In addition, to measure progress toward the goal to slow the emergence of resistant bacteria and prevent the spread of resistant infections, the National Action Plan set a significant outcome of reducing inappropriate antibiotic use by 50 percent in outpatient settings and by 20 percent in inpatient settings by 2020—a performance target that requires the collection, analysis, and reporting of antibiotic use data.

coordinators said, in response to our inquiries during this review, that the task force intends to identify agencies' plans for addressing barriers in the Year 5 progress report to be published in fall 2020. However, the coordinators also stated that the progress reports to date have not identified plans to address barriers largely because the task force focused on reporting the agencies' accomplishments in implementing the National Action Plan. Until the CARB Task Force identifies and reports on agencies' plans to address barriers related to the expansion of antibiotic stewardship programs and the collection of antibiotic use data across health care settings to the extent feasible, the federal government will not have reasonable assurance that it is fully implementing the National Action Plan and addressing antibiotic resistance.

Antibiotic Stewardship Training May Have Limited Success in Improving Prescribing Behavior

While training is recognized as one component of an antibiotic stewardship program, such training may have limited success in improving antibiotic prescribing behavior, and federal agencies indicate that it is challenging to evaluate the training's effectiveness. CDC officials and experts say that inappropriate antibiotic use could be improved through stewardship training, but it is challenging because antibiotic prescribing behavior is driven by multiple factors and can be difficult to change. For example, a PACCARB report stated that prescribers often feel pressure to prescribe antibiotics—even when antibiotics may not be warranted—because of their perception that a patient is demanding such a prescription, or a patient's actual demand.[193] In addition, CDC notes that antibiotics are frequently prescribed for respiratory conditions most commonly caused by viruses such as the common cold, against which antibiotics are ineffective. Other factors that drive antibiotic prescribing behavior, as cited by experts, include habit, which may stem from what physicians and other prescribers learn during their residencies or observe in the workplace; the time it takes

[193] Presidential Advisory Council on Combating Antibiotic-Resistant Bacteria, *Key Strategies to Enhance Infection Prevention and Antibiotic Stewardship: Report with Recommendations for Human and Animal Health* (Washington, D.C.: September 2018).

to explain to a patient why an antibiotic is inappropriate; and "decision fatigue" caused by tiredness or hunger. (See Table 4 for examples of factors that drive or deter antibiotic prescribing behavior).

Table 4. Examples of factors that drive or deter antibiotic prescribing behavior

Factors that drive antibiotic prescribing	Factors that deter antibiotic prescribing
• Perception that patients want antibiotics • Fear that patients will not be satisfied with their care without a prescription, leading patients to give the provider poor ratings or reviews • Perception that it is easier and quicker to prescribe antibiotics than to explain to patients why they are unnecessary • No billing code or reimbursement for "stewardship" • Habit • Decision fatigue caused by tiredness or hunger • Worry about serious complications (without antibiotics) and a "just to be safe" mentality • Lack of financial incentives to use diagnostic tests (e.g., to determine if an infection is bacterial vs. viral) • Reluctance to wait to see if a patient's symptoms continue before prescribing	• Risks of adverse reactions • Risks of drug interactions • Recognition of the need for antibiotic stewardship • Desire to deter low-value care[a] • Desire to decrease unnecessary health care spending • Preference to follow prescribing guidelines • Comparison with peers of antibiotic prescriptions data • Use of a "pre-commitment" poster/letter to patients that the medical practice is dedicated to appropriate use of antibiotics • "Accountable justification"—a pop-up window that appears on electronic health record systems, alerting providers that antibiotics are not generally indicated for the diagnosis and asking the provider to enter a reason to justify the prescription

Source: GAO summary of information from A. Mehrotra and J. A. Linder, "Tipping the Balance Toward Fewer Antibiotics," *JAMA Internal Medicine*, vol. 176, no. 11 (2016): pp. 1649-1650; a Presidential Advisory Council on Combating Antibiotic-Resistant Bacteria September 2017 meeting summary; a GAO expert meeting in September 2018; and experts we interviewed. | GAO-20-341.

Note: This list of factors is not intended to be comprehensive but, rather, represents a summary of factors presented by the information we reviewed.

[a]In health care, low-value care refers to treatments, tests, and procedures that are either wasteful or ineffective, such as an antibiotic prescription for a viral infection.

Nevertheless, federal agencies plan to evaluate the effectiveness of their antibiotic stewardship training programs to some extent, although the National Action Plan does not require the agencies to do so. For example,

CDC officials told us that their online training course for various types of clinicians allows participants to fill out an evaluation that includes questions about whether the participant will be able to apply knowledge gained from the course, which the agency will use to refine and update the course.[194] In addition, for the antibiotic stewardship training for nursing homes that CDC and CMS jointly developed, CDC officials told us that participants will be asked 6 months after the training whether participants implemented stewardship practices—and whether there have been reductions in antibiotic use—as a result of the training. However, CMS, DOD, and VA officials noted that it is difficult to isolate and measure the effectiveness of antibiotic stewardship training specifically on antibiotic prescribing behavior—compared to other, concurrent federal efforts, such as requirements and guidance to promote appropriate antibiotic use. For example, DOD officials told us that their department has looked at antibiotic use data from DOD health care facilities as a "surrogate" to evaluate whether antibiotic stewardship in general has been effective— but noted that is an imperfect measure since there are many factors that affect antibiotic prescribing behavior, and training is only one of several interventions aimed at reducing inappropriate antibiotic use.

CONCLUSION

Antibiotic resistance has been characterized as one of the greatest public health threats the world faces. A concerted effort involving coordination of multiple stakeholders and countries and across health fields is critical to helping ensure that bacterial infections remain treatable. Steps by federal agencies to expand surveillance, facilitate the development and use of new diagnostic tests, fund R&D for the development of new treatments, and issue requirements and guidance for antibiotic stewardship programs are important efforts toward addressing the problem of antibiotic resistance and implementing the National Action Plan.

194 CDC officials also told us that the agency piloted the online training course with target audiences prior to its release in order to improve the course.

Significant challenges to conducting surveillance remain. For example, CDC has not determined the participation rates or appropriate distribution of participating hospitals needed by the voluntary antibiotic-resistance reporting option to achieve CDC's goal of conducting regional and national assessments of resistance. By taking steps to determine the participation rates and distribution needed for this option, CDC would have more reasonable assurance that it can achieve its goal. CDC classified gonorrhea as one of the most urgent resistant threats in the nation, but collects limited specimens—representing an estimated 1 to 2 percent of the reported cases in the United States—for GISP, its primary surveillance system for resistant gonorrhea. However, CDC has not fully evaluated the representativeness of the trends identified by this surveillance system. By evaluating GISP to ensure that it includes measures of its representativeness, such as comparing the trends in the sample population with those in the overall U.S. population, using specially designed studies if needed, CDC would have better assurance that the trends detected in GISP accurately reflect the characteristics of the health-related outcome the system is designed to monitor. Further, neither the 2013 nor the 2019 Threats Reports provided quantitative measures of uncertainty for CDC's estimates of morbidity and mortality resulting from antibiotic-resistant infections. Providing such measures, such as standard errors or confidence intervals, as appropriate, in its Threats Reports would help CDC and others compare information within and across reporting efforts, and draw appropriate conclusions about the characteristics of antibiotic resistance in the United States, including limitations associated with reported findings and conclusions. Finally, there has been a 6-year interval between CDC's reports on antibiotic resistance threats. By developing a plan for more frequent dissemination of consolidated reporting on priority pathogens at regular intervals, CDC would have more timely trend data and other information necessary for users of the data, including policymakers, to prioritize, plan, implement, and evaluate public health actions to address antibiotic resistance.

HHS has funded some studies to assess whether certain tests for antibiotic resistance lead to improved clinical outcomes, including more effective treatment for patients or more judicious use of antibiotics. However, HHS agencies that are in a position to conduct or fund such studies have not identified leadership, roles, and responsibilities to help further such efforts. By taking steps to identify leadership, roles, and responsibilities, agencies could more effectively address the need for clinical outcomes studies, potentially increasing test use, improving patient care, and enhancing stewardship efforts. In addition, for its part, FDA has not regularly monitored tests for antibiotic resistance to assess breakpoint updates or evaluated any effects of using tests for antibiotic resistance with out-of-date breakpoints. By regularly monitoring and evaluating FDA-authorized tests that rely on breakpoints, FDA would be able to determine whether test manufacturers are updating breakpoints as needed and help ensure that patient care and infection control efforts are effective.

While government push incentives to support antibiotic R&D have been helpful, experts and antibiotic developers have indicated that push incentives alone are not sufficient to sustain antibiotic development. PACCARB, TATFAR, and other experts have called for additional postmarket pull incentives to increase the antibiotic pipeline, but HHS does not have a strategy for doing so. Developing a strategic framework that outlines key design elements of new incentives would be a first step toward identifying potential authorities and resources that may be needed and determining agency roles for implementation and oversight of the incentives. Until such incentives are developed, more drug companies may exit the antibiotic development sector, and the pipeline of new treatments may continue to decrease.

Finally, in its progress reports covering the first four years of the National Action Plan's implementation, the CARB Task Force did not identify plans, as required by the Executive Order, to address barriers that agencies face in fully implementing the National Action Plan, such as expanding (1) a CDC program designed to strengthen the U.S. response to resistant gonorrhea; (2) antibiotic stewardship programs across health care settings; and (3) antibiotic use data collection, to the extent feasible. Without identifying plans to address these and other challenges, the federal government cannot assure that the country is prepared to overcome the urgent health consequences of antibiotic resistance. Until the CARB Task Force, which is coordinated by HHS officials, identifies and reports on agencies' plans to address barriers preventing full implementation of the National Action Plan, the federal government will not have reasonable assurance that it is fully implementing the National Action Plan and addressing antibiotic resistance.

RECOMMENDATIONS FOR EXECUTIVE ACTION

We are making a total of eight recommendations, including four to CDC, three to HHS, and one to FDA. Specifically:

The Director of CDC should take steps to determine participation rates and distribution needed in the AR Option of the National Healthcare Safety Network for conducting regional and national assessments of antibiotic resistance of public health importance. (Recommendation 1)

The Director of CDC should ensure that CDC's evaluation of its surveillance system for antibiotic-resistant gonorrhea includes measures of its representativeness, such as comparison of the trends in the sample population with those in the overall U.S. population, using specially designed studies if needed. (Recommendation 2)

The Director of CDC should provide information on uncertainties for antibiotic resistance estimates in its consolidated Threats Reports, including standard errors or confidence intervals, as appropriate. (Recommendation 3)

The Director of CDC should develop a plan for timely, consolidated reports of antibiotic resistance in priority pathogens at regular intervals. (Recommendation 4)

The Secretary of HHS should identify leadership and clarify roles and responsibilities among HHS agencies to assess the clinical outcomes of diagnostic testing for identifying antibiotic-resistant bacteria. (Recommendation 5)

The Commissioner of FDA should direct the Center for Devices and Radiological Health to conduct additional monitoring and evaluation of the status of FDA-authorized tests that rely on breakpoints, on a regular basis, to determine whether test manufacturers are updating breakpoints, seeking additional resources as needed. (Recommendation 6)

The Secretary of HHS should develop a strategic framework to further incentivize the development of new treatments for antibiotic-resistant infections, including through the use of postmarket financial incentives, and, if appropriate, make recommendations to Congress for necessary authority. (Recommendation 7)

The Secretary of HHS should direct the CARB Task Force to include in its annual updates to the President plans for addressing any barriers preventing full implementation of the National Action Plan and, as appropriate, make recommendations for new or modified actions.

Specifically, the CARB Task Force should identify plans to address barriers, such as those related to expanding (1) a CDC program designed to strengthen the U.S. response to resistant gonorrhea; (2) antibiotic stewardship programs across health care settings; and (3) antibiotic use data collection across health care settings, to the extent feasible. (Recommendation 8).

AGENCY COMMENTS AND OUR EVALUATION

We provided a draft of this report to DOD, VA, and HHS for review and comment. DOD and VA did not provide formal comments but generally agreed with our report.

In its comments, reproduced in appendix V, HHS generally concurred with our findings and seven of our recommendations, and did not concur with one of our recommendations, as discussed below. HHS identified several actions it intends to take to address our recommendations. DOD and HHS also provided technical comments, which we incorporated as appropriate.

In response to our first recommendation, HHS concurred and CDC stated it is working with public health partners to promote the voluntary use of the AR Option, providing technical support to states that may be considering a state or local mandate to require AR and AU reporting, and developing pilot programs to assess AR Option data and other data sources for certain types of antibiotic resistance. While these actions are helpful, we believe taking additional steps, such as determining goals for participation rates and distribution for AR Option reporting, would give CDC more reasonable assurance that hit can conduct regional and national assessments of resistance.

In response to our second recommendation, HHS concurred and CDC stated it is taking additional efforts to examine the representativeness of data collected through its primary surveillance system for resistant gonorrhea, including working to develop laboratory methods to reduce dependence on cultured isolates. CDC stated that steps to refine and improve collection of resistant gonorrhea data require additional resources. We believe that CDC requesting such resources would help ensure that such data are representative of the overall U.S. population.

HHS generally concurred with our third recommendation. CDC stated it feels that it is critical to publish the data after peer review and then plans to link the publications back to online resources of the 2019 Threats Report. We believe that peer-reviewed publication is important, but it is also important for CDC to take additional steps to establish and report uncertainties for the national estimates or summary data that would help CDC and others draw appropriate conclusions about the characteristics of antibiotic resistance in the United States.

In response to our fourth recommendation, HHS concurred and CDC stated it has plans to update its enterprise-wide AR Threats Report every three years, and that it also issues regular reports on specific groups of pathogens.

In response to our fifth recommendation, HHS concurred and stated that the CARB Task Force leadership will work with relevant HHS agencies to clarify roles and responsibilities and identify leadership, if appropriate, for supporting research on clinical outcomes delated to diagnostic tests.

HHS concurred with our sixth recommendation, and FDA concurred with conducting additional monitoring and evaluation of tests relying on breakpoints when FDA identifies or recognizes new breakpoints. FDA stated that it has taken major steps to help address challenges associated with updating such tests to reflect the most current breakpoints. We believe that in addition to these steps, monitoring and evaluation of current FDA-authorized tests that may still be using out-of- date breakpoints will enhance FDA's ability to provide assurance that patient care and infection control efforts are effective.

HHS did not concur with our seventh recommendation that HHS should develop a strategic framework to further incentivize the development of new treatments for antibiotic-resistant infections, including through the use of postmarket financial incentives. HHS noted that, while it agrees that additional incentives are needed to address the limited pipeline for novel and innovative treatments to combat antibiotic resistance, it is still conducting analyses to understand whether postmarket incentives should be included as a component of its forthcoming strategic framework to further incentivize the development of new treatments. However, HHS did not specify when its framework would be released. We support HHS's efforts to develop such a framework, as this is a complex issue with multiple factors to consider. However, we believe our recommendation is still warranted. Antibiotic resistance is one of the greatest global public health threats, and experts, including the WHO, have warned that the pipeline of new antibiotics in development is insufficient to combat the threat. Without an adequate arsenal of treatments, we are likely to see increasing mortality caused by these deadly infections. As we reported, experts, advisory groups, federal officials, and antibiotic developers have all called for additional postmarket incentives to reinvigorate the pipeline of antibiotics under development. The current significant federal investment in push incentives to support antibiotic R&D is helpful but will ultimately be ineffective if companies receiving this investment are unable to sustain their business once their treatment reaches the market. Therefore, we maintain that it is important that HHS not delay the development of a strategic framework that includes postmarket incentives, which is just an initial step toward the creation of these incentives. Until additional postmarket incentives are developed, more drug companies may exit the antibiotic development sector, and the pipeline of new treatments for antibiotic-resistant infections may continue to decrease.

In response to our eighth recommendation, HHS concurred and stated that beginning in 2020 and continuing annually thereafter, the CARB Task Force's progress reports will include discussion of any barriers preventing full implementation of the National Action Plan, including, as appropriate, barriers that GAO has identified. We emphasize that the CARB Task Force should also identify plans to address such barriers—and, as appropriate, make recommendations for new or modified actions—in future progress reports, in accordance with Executive Order No. 13676.

As agreed with your offices, unless you publicly announce the contents of this report earlier, we plan no further distribution until 30 days from the report date. At that time, we will send copies to appropriate congressional committees; the Secretaries of DOD, HHS, and VA; and other interested parties. In addition, the report will be available at no charge on the GAO website at http://www.gao.gov.

If you or your staff have any questions about this report, please contact us at (202) 512-6888 or personst@gao.gov, or (202) 512-7114 or deniganmacauleym@gao.gov. Contact points for our Offices of Congressional Relations and Public Affairs may be found on the last page of this report. GAO staff who made key contributions to this report are listed in appendix VI.

Timothy M. Persons, PhD,
Chief Scientist and Managing Director Science,
Technology Assessments, and Analytics

Mary Denigan-Macauley, PhD, Director
Health Care

Appendix I: Objectives, Scope, and Methodology

This report examines: (1) the Centers for Disease Control and Prevention's (CDC) efforts to address surveillance of antibiotic resistance and any challenges to these efforts; (2) federal efforts to advance the development and use of diagnostic tests for identification and characterization of resistant bacteria and to address barriers to the development of diagnostic tests; (3) challenges to developing new treatments for antibiotic-resistant infections and federal efforts to address the challenges; and (4) federal efforts to promote the appropriate use of antibiotics and any challenges that remain.

We focused our review primarily on agency actions since 2015, when the *National Action Plan for Combating Antibiotic-Resistant Bacteria* (National Action Plan) was published.[195] We also focused our review on human health, as we have reported on federal efforts to address the use of antibiotics in food animals and recommended actions to improve these efforts for more than 20 years.[196] Additionally, we focused our review on antibiotic-resistant bacteria. We generally excluded federal efforts related to infection prevention and control in human health care, on which we have previously reported.[197]

To address all four objectives, we reviewed relevant agency reports and documents, such as CDC's report, *Antibiotic Resistance Threats in the United States, 2013* (2013 Threats Report); conducted interviews with

---

[195] The White House, *National Action Plan for Combating Antibiotic-Resistant Bacteria* (Washington, D.C.: March 2015).

[196] GAO, *Antibiotic Resistance: More Information Needed to Oversee Use of Medically Important Drugs in Food Animals*, GAO-17-192 (Washington, D.C.: Mar. 2, 2017); *Antibiotic Resistance: Agencies Have Made Limited Progress Addressing Antibiotic Use in Animals*, GAO-11-801 (Washington, D.C.: Sept. 7, 2011); *Antibiotic Resistance: Federal Agencies Need to Better Focus Efforts to Address Risk to Humans from Antibiotic Use in Animals*, GAO-04-490 (Washington, D.C.: Apr. 22, 2004); and *Food Safety: The Agricultural Use of Antibiotics and Its Implications for Human Health*, GAO/RCED-99-74 (Washington, D.C.: Apr. 28, 1999).

[197] GAO, *Health-Care-Associated Infections in Hospitals: Leadership Needed from HHS to Prioritize Prevention Practices and Improve Data on These Infections*, GAO-08-283 (Washington, D.C.: Mar. 31, 2008).

officials from federal agencies, experts, and stakeholder organizations; and we reviewed relevant literature, policy papers, and GAO reports.[198] We interviewed officials from federal agencies responsible for implementing the aspects of the National Action Plan related to our research objectives: the Department of Health and Human Services' (HHS) Office of the Assistant Secretary for Planning and Evaluation, the Biomedical Advanced Research and Development Authority (BARDA), CDC, the Centers for Medicare & Medicaid Services (CMS), the Food and Drug Administration (FDA), the National Institutes of Health (NIH), and the Office of Global Affairs; as well as the Department of Defense (DOD) and the Department of Veterans Affairs. We also interviewed experts and representatives from organizations involved in public health and epidemiology, infectious diseases and microbiology, antibiotic research and development (R&D), antibiotic stewardship, and other issues relating to antibiotic resistance. Because antibiotic resistance is a global problem, we also interviewed officials from the World Health Organization (WHO), the European Centre for Disease Prevention and Control, the European Medicines Agency, the Wellcome Trust, Public Health England, and the Surveillance and Epidemiology of Drug-Resistant Infections Consortium about various aspects of our review; and we reviewed relevant documents from these entities. We identified experts and organizations through literature and other documents we reviewed and through referrals from agency officials and other experts we interviewed.[199] In addition, we attended several meetings and reviewed summaries of meetings held by the Presidential Advisory Council on Combating Antibiotic-Resistant Bacteria (PACCARB). Furthermore, we attended two conferences related to antibiotic resistance: the World Anti-Microbial Resistance Congress and the Gordon Research Conference on chemical and biological threat defense, the latter of which had a session devoted to antibiotics and antibiotic resistance. For each of our objectives, we identified and reported on actions taken by federal agencies and key challenges that the agencies

---

[198] Department of Health and Human Services, Centers for Disease Control and Prevention, *Antibiotic Resistance Threats in the United States, 2013* (Atlanta, Ga.: Apr. 23, 2013).

[199] We assessed the studies we cite in this report for their methodological limitations and determined that they were sufficiently reliable for our purposes.

face in addressing antibiotic resistance. We evaluated the actions taken by federal agencies against relevant criteria, as applicable.

In addition, in September 2018, we convened a meeting of experts in antibiotic resistance epidemiology, diagnostic testing, antibiotic development, and antibiotic stewardship. This meeting of experts was planned and convened with the assistance of the National Academy of Sciences to better ensure that a breadth of expertise was brought to bear in its preparation; however, all final decisions regarding meeting substance and expert participation are the responsibility of GAO. Any conclusions and recommendations in GAO reports are solely those of the GAO. The Board on Population Health and Public Health Practice within the National Academy of Sciences solicited expert nominations from academia, public health laboratories, industry, and other organizations working in topics relating to antibiotic resistance. From their list of 51 nominees, and additional nominees we independently identified, we convened a meeting of 18 experts selected for their knowledge and expertise related to antibiotic resistance epidemiology, diagnostic testing, antibiotic development, and antibiotic stewardship. Eleven of the 18 experts who participated in our meeting also reviewed and provided comments on a draft of our report. We refer to such experts in this report as "experts at our meeting;" appendix II contains a list of the expert participants.

To examine CDC's efforts to address surveillance for antibiotic resistance and any challenges to these efforts, we reviewed documentation and conducted interviews with agency officials and other key stakeholders on each of the surveillance systems across CDC that track antibiotic resistance and reviewed CDC's 2013 Threats Report and CDC's *Antibiotic Resistance Threats in the United States, 2019* data.[200] We further focused our review on the 17 priority disease-causing bacteria listed in CDC's 2013 Threats Report. The CDC surveillance systems included:

- Antibiotic Resistance Laboratory Network
- Emerging Infections Program (EIP)

---

[200] Department of Health and Human Services, Centers for Disease Control and Prevention, *Antibiotic Resistance Threats in the United States, 2019* (Atlanta, Ga.: Nov. 13, 2019).

- Gonococcal Isolate Surveillance Program (GISP)
- National Antimicrobial Resistance Monitoring System (NARMS)
- National Healthcare Safety Network (NHSN)
- National Notifiable Diseases Surveillance System
- National Tuberculosis Surveillance System

For NHSN, we also assessed health care facility participation data by state and territory. We assessed the reliability of these data by reviewing them for any outliers or anomalies and by inquiring with agency officials about their source and any known reliability issues. We determined that these data were sufficiently reliable for assessing facility participation rates by U.S. state and territory. Stakeholder organizations we interviewed represented state and territorial epidemiologists and other public health officials (the Council of State and Territorial Epidemiologists and the Association of State and Territorial Health Officials) and an international consortium to address challenges in surveillance of antibiotic resistance (the Surveillance and Epidemiology of Drug-resistant Infections Consortium). We also reviewed reports on antibiotic resistance surveillance challenges from the Public Health Informatics Task Force and the Antibiotic Resistance Surveillance Task Force. We also reviewed documents from WHO's global surveillance system and interviewed WHO and CDC officials to identify challenges that limit CDC's ability to assess threats from abroad. We evaluated challenges and steps CDC has taken against CDC's "Updated Guidelines for Evaluating Public Health Surveillance Systems;" *Standards for Internal Control in the Federal Government*; prior GAO work; the Government Performance and Results Act of 1993 (GPRA) and the GPRA Modernization Act of 2010; the Office of Management and Budget Circular No. A-11 and *Standards and Guidelines for Statistical Surveys*; relevant National Action Plan objectives, aims, and milestones; and Executive Order No. 13676, September 2014.[201]

---

[201] Department of Health and Human Services, Centers for Disease Control and Prevention, "Updated Guidelines for Evaluating Public Health Surveillance Systems," *Morbidity and Mortality Weekly Report*, vol. 50, no. RR-13 (2001); GAO, *Standards for Internal Control*

To examine federal efforts to advance the development and use of diagnostic tests, we also interviewed representatives from a nongeneralizable selection of six diagnostic test manufacturers to identify challenges they face in developing tests for antibiotic resistance and challenges in increasing user adoption of their tests. We further focused our review on the 17 priority disease-causing bacteria listed in CDC's 2013 Threats Report. The six manufacturers we interviewed were Accelerate Diagnostics, Beckman Coulter, BioFire and its parent company, BioMerieux, Bruker, Cepheid, and Roche Diagnostics. We identified these manufacturers by compiling a list based on previous work we conducted, interviews with select experts, and internet search. We selected six manufacturers that were identified by more than one source while encompassing different types of tests (culture and genotypic). We limited our scope to FDA-authorized tests—that is, tests that have been reviewed and cleared by FDA for marketing in the United States—that can identify resistance in at least one type of bacteria categorized as priority bacteria in CDC's 2013 Threats Report.[202] Some of these tests are called antibiotic susceptibility tests, but we refer to the entire class of such tests as "tests." We included in our scope tests that can differentiate between viral and bacterial infection because these types of tests are included in the National Action Plan. We evaluated the actions taken by federal agencies against the *Standards for Internal Control in the Federal Government*, relevant National Action Plan objectives, aims, and milestones under Goal 3, and relevant sections in the PACCARB *Recommendations for Incentivizing the*

---

in the Federal Government, GAO-14-704G (Washington, D.C.: Sept. 10, 2014); GAO, *Performance.gov: Long-Term Strategy Needed to Improve Website Usability*, GAO-16-693 (Washington, D.C.: Aug. 30, 2016); Government Performance and Results Act of 1993, Pub. L. No. 103-62, 107 Stat. 285 (Aug. 3, 1993); GPRA Modernization Act of 2010, Pub. L. No. 111-352, 124 Stat. 3866 (Jan. 4, 2011); Office of Management and Budget, Circular No. A-11, *Preparation, Submission, and Execution of the Budget*, Pt. 6, § 200.22 (Washington, D.C.: June 28, 2019) and Office of Management and Budget, *Standards and Guidelines for Statistical Surveys* (September 2006); and Exec. Order No. 13676, *Combating Antibiotic-Resistant Bacteria*, 79 Fed. Reg. 56931 (Sept. 23, 2014).

[202] FDA officials told us that the tests under GAO review are Class II devices and thus reviewed as 510(k) or de novo applications and are appropriately referred to as cleared when granted marketing authorization. Other types of diagnostic tests, such as laboratory- developed tests, are not within the scope of this report. FDA describes laboratory- developed tests as tests designed, manufactured, and used within a single laboratory.

Development of Vaccines, Diagnostics, and Therapeutics to Combat Antibiotic Resistance. We also evaluated federal agency actions against the "leadership" and "clarity of roles and responsibilities" leading practices from GAO's *Managing for Results: Key Considerations for Implementing Interagency Collaborative Mechanisms*.[203] We focused on these key practices when there was a lack of specifically assigned roles in either the National Action Plan or the PACCARB report for key activities.

To identify challenges to developing new treatments for antibiotic-resistant infections and examine federal efforts to address these challenges, we also interviewed 11 randomly selected companies that conduct research and development on new treatments for bacterial infections.[204] We included companies that are researching or developing both traditional antibiotics and alternatives to antibiotics—which we call "nontraditional products" in this report—and we included companies that had and had not received funding from the Combating Antibiotic- Resistant Bacteria Biopharmaceutical Accelerator (CARB-X) and companies that do and do not have existing FDA-approved drugs on the market. We asked company representatives about challenges in developing new antibiotics they have identified, support they may have received from federal agencies, how effective the support has been to them, and their views on additional incentives that would promote the development of new antibiotics. We also interviewed experts on the topic of antibiotic development and industry stakeholders, specifically The Pew Charitable Trusts and the Biotechnology Innovation Organization. We interviewed federal officials from BARDA, CMS, DOD, FDA, and NIH to learn about their programs and actions to support the development of treatments for antibiotic-resistant infections and requested information about funding for antibiotic R&D from BARDA, DOD, and NIH. We included relevant agency actions

---

[203] GAO-14-704G; *National Action Plan*, Goal 3: Advance Development and Use of Rapid and Innovative Diagnostic Tests for Identification and Characterization of Resistant Bacteria; Presidential Advisory Council on Combating Antibiotic-Resistant Bacteria, *Recommendations for Incentivizing the Development of Vaccines, Diagnostics, and Therapeutics to Combat Antibiotic Resistance* (Washington, D.C.: September 2017); and GAO, *Managing for Results: Key Considerations for Implementing Interagency Collaborative Mechanisms,* GAO-12-1022 (Washington, D.C.: Sept. 27, 2012).

[204] We initially selected 12 companies to interview, but one company declined.

that began before the National Action Plan was issued in 2015 if they continued after 2015. Finally, we reviewed literature related to antibiotic development and reports about antibiotic pull incentives written by health policy advisory groups, including the PACCARB, the Transatlantic Taskforce on Antimicrobial Resistance (TATFAR), the DRIVE-AB project, and the Duke Margolis Center for Health Policy.[205] We evaluated the actions taken by federal agencies to help address the challenges to developing new treatments against experts' and advisory groups' views on additional actions needed and against the principles related to developing strategies outlined in GPRA and the GPRA Modernization Act of 2010.[206] We did not assess challenges to developing products designed to prevent infections, such as vaccines, nor federal actions related to these types of products.

To examine federal agency efforts to promote the appropriate use of antibiotics and any challenges that remain, we also analyzed CMS data and related documentation on the quality measures and improvement activities related to antibiotics as part of CMS's Merit-based Incentive Payment System (MIPS) in 2017. Specifically, we identified CMS's antibiotics-related quality measures and improvement activities by conducting a search for the words "antibiotic," "antimicrobial," "bacteria," "resistance," and "resistant" on CMS's MIPS website.[207] We then reviewed CMS's data on the number of MIPS-eligible clinicians who selected and reported on these measures and activities in 2017, the most recently available data. In 2017, there were nine MIPS quality measures related to antibiotics, as follows:

1) acute otitis externa: systemic antimicrobial therapy - avoidance of inappropriate use;
2) adult sinusitis: antibiotic prescribed for acute sinusitis (overuse);

---

[205] DRIVE-AB is the short name for the "Driving Re-investment in R&D and Responsible Antibiotic Use."
[206] Pub. L. No. 103-62, 107 Stat. 285 and Pub. L. No. 111-352, 124 Stat. 3866.
[207] Centers for Medicare & Medicaid Services, "Quality Payment Program: Explore Measures & Activities," accessed November 6, 2019, https://qpp.cms.gov/mips/explore- measures.

3) adult sinusitis: appropriate choice of antibiotic: amoxicillin with or without Clavulanate prescribed for patients with acute bacterial sinusitis (appropriate use);
4) appropriate testing for children with pharyngitis;
5) appropriate treatment for children with upper respiratory infection;
6) appropriate treatment of Methicillin-sensitive Staphylococcus aureus bacteremia;
7) avoidance of antibiotic treatment in adults with acute bronchitis;
8) perioperative care: selection of prophylactic antibiotic – first- or second-generation Cephalosporin; and
9) total knee replacement: preoperative antibiotic infusion with proximal tourniquet.

In addition, there was one MIPS improvement activity related to antibiotics in 2017: implementation of antibiotic stewardship program. We reviewed the MIPS data for any obvious outliers or anomalies, and we determined that these data were sufficiently reliable for reporting the number of clinicians who reported implementing these quality measures and improvement activities. In addition, we reviewed aggregated data from CDC on the total number of eligible U.S. hospitals voluntarily reporting their antibiotic use data to a CDC system (the NHSN's Antimicrobial Use Option); we then calculated the percentage of eligible hospitals reporting such data as of January 1, 2020. We assessed the reliability of the aggregated data by reviewing them for any obvious errors or missing data totals and inquiring with CDC officials about their source and any known reliability issues. We determined that these data were sufficiently reliable for reporting hospital participation rates in the system. We also reviewed selected articles on antibiotic use and stewardship— compiled from a variety of sources, including CDC documents and experts we interviewed—published in literature. In addition, we interviewed experts on antibiotic use and stewardship, including representatives from PACCARB, Emory University's School of Medicine, the University of Minnesota's Center for Infectious Disease Research and Policy, The Joint Commission, the Society of Infectious Diseases Pharmacists, The Pew

Charitable Trusts, and the Association for Professionals in Infection Control and Epidemiology. We evaluated federal efforts and challenges against relevant National Action Plan objectives and milestones and Executive Order No. 13676. We focused on antibiotic use in the United States, rather than global antibiotic use.

We conducted this performance audit from February 2018 to March 2020 in accordance with generally accepted government auditing standards. Those standards require that we plan and perform the audit to obtain sufficient, appropriate evidence to provide a reasonable basis for our findings and conclusions based on our audit objectives. We believe that the evidence obtained provides a reasonable basis for our findings and conclusions based on our audit objectives.

APPENDIX II: EXPERT MEETING PARTICIPANT LIST

We collaborated with the National Academy of Sciences to convene a two-day meeting of experts to inform our work on federal efforts to address antibiotic resistance; the meeting was held on September 17 and 18, 2018. The experts who participated in this meeting are listed below. Many of these experts gave us additional assistance throughout our work, including by providing additional technical expertise and answering questions, and 10 of these experts reviewed and provided comments on our draft report for technical accuracy.

Helen W. Boucher, MD, FACP, FIDSA
Director, Infectious Diseases Fellowship Program
Director, Heart Transplant and Ventricular Assist Device Infectious Diseases Program
Professor, Tufts University School of Medicine, Division of Geographic Medicine and Infectious Diseases, Tufts Medical Center
Boston, MA

Kate Cook
Executive Vice President
Drugs and Biological Products
Greenleaf Health, Inc.
Washington, DC

Stan Deresinski, MD, FACP, FIDSA
Clinical Professor of Medicine
Infectious Diseases
Stanford University School of Medicine
Stanford, CA

Mary Jane Ferraro, PhD, MPH
Pathologist, Subspecialty Co-Head, Microbiology Laboratory, Massachusetts General Hospital
Professor of Pathology & Professor of Medicine, Harvard Medical School
Boston, MA

Susan Huang, MD, MPH.
Professor, Infectious Disease, Medical Director
Epidemiology and Infection Prevention
School of Medicine, University of California, Irvine
Irvine, CA

Romney Humphries, PhD, D(ABMM), M(ASCP)
Chief Scientific Officer
Accelerate Diagnostics
Tucson, AZ

Timothy Jinks, PhD
Head of Drug Resistant Infections Priority Program
Wellcome Trust
London, UK

Marion A. Kainer, MD, MPH., FRACP, FSHEA
Director, Healthcare Associated Infections and Antimicrobial Resistance Program
Tennessee Department of Health
Nashville, TN

Lonnie King, DVM, MS, MPA., Diplomate ACVPM
Professor and Special Assistant to the Provost
Department of Veterinary Preventive Medicine, Office of Academic Affairs
The Ohio State University
Columbus, OH

Colleen S. Kraft, MD, MSc
Associate Professor
Pathology & Laboratory Medicine
Emory University School of Medicine
Atlanta, GA

Ramanan Laxminarayan, PhD, MPH.
Senior Research Scholar and Lecturer, Princeton University
Princeton, NJ
Director and Senior Fellow, Center for Disease Dynamics
Washington, DC
Co-founder, HealthCubed
New Delhi, India

Mary Lou Manning, PhD, CRNP, CIC, FAAN, FAPIC
Professor
Thomas Jefferson University
Philadelphia, PA

Mark Miller, MD, MSc, FRCP(C)
Chief Medical Officer
bioMérieux
Marcy L'Etoile, France

Robert A. Myers, PhD
Director
Laboratories Administration
Maryland Department of Health
Baltimore, MD

John H. Rex, MD, FACP
Chief Medical Officer and Director
F2G, Ltd.
Boston, MA

Marc Sprenger, MD, PhD
Director, Antimicrobial Resistance
Office of the Director-General
World Health Organization Geneva, Switzerland

John M. Stelling, MD, MPH
Co-Director, WHO Collaborating Center for Surveillance of Antimicrobial Resistance
Instructor, Department of Medicine
Brigham and Women's Hospital
Boston, MA

Barrett Thornhill, JD
Executive Director
Antimicrobial Innovation Alliance
Washington, DC

Appendix III: Additional Examples of Federal Efforts to Support Antibiotic Research and Development

This appendix contains additional examples of efforts by agencies within the Departments of Health and Human Services, Defense, and Energy to provide support for antibiotic research and development beyond those mentioned in the report. These examples do not comprise the full extent of agencies' efforts.

Table 5. Additional Examples of Federal Efforts to Support Antibiotic Research and Development

Agency	Efforts
Department of Health and Human Services	
National Institutes of Health (NIH)	The NIH's National Institute of Allergy and Infectious Diseases (NIAID) offers three distinct mechanisms to support research and development of treatments for antibiotic-resistant infections: 1) Grants for basic and applied research: for example, research to better understand how bacteria develop resistance and the interactions between human microbial communities and drug-resistant bacteria 2) Product development contracts: up to 5-year contracts to support the development of new treatments, vaccines, or diagnostic products. Contracts provide companies with funding and access to NIAID experts to decrease product development risk. 3) Pre-clinical and clinical services: research services provided free-of-charge to external researchers; for example, in-vitro assessment of therapeutic candidates' antimicrobial activity and services to support clinical trials for new drugs.
	NIAID's Antibacterial Resistance Leadership Group provides extramural researchers with funding for clinical research studies focused on: infections caused by gram-positive and gram- negative bacteria, and diagnostics development. As of June 2019, the Antibacterial Resistance Leadership Group has initiated more than 40 studies at 130 clinical trial sites.
	Through its Centers of Excellence for Translational Research, NIAID funds research centers throughout the United States to research and develop new or improved medical countermeasures for emerging infectious diseases. Five of the 14 awarded centers are engaged in researching treatments for antibiotic resistance.

Agency	Efforts
Department of Health and Human Services	
	For example, one center at Harvard University Medical School is researching how best to target the bacterial cell envelope for several types of antibiotic-resistant pathogens and aims to develop at least one new vaccine and three to five antibacterial compounds that demonstrate efficacy in animal models
	NIH officials told us NIAID convenes scientific workshops—10 since 2015—to facilitate the development of approaches to challenging scientific questions and promote collaboration among federal agencies and academic and industry researchers.
	NIAID collaborates with other countries to advance antibiotic-resistance research worldwide. For example, as a member of the Transatlantic Taskforce on Antimicrobial Resistance (TATFAR), NIAID coordinates and aligns international research activities and organizes international workshops. A result of the TATFAR collaboration is the launch of a partnership between U.S. and European entities to align clinical trial networks to increase access to patients for clinical trials. NIAID has also partnered with the Chinese government to advance antibiotic resistance research activities and clinical trials.
	The National Center for Biotechnology Information maintains the National Database of Antibiotic-Resistant Organisms in partnership with several other federal agencies, including the Centers for Disease Control and Prevention, the Food and Drug Administration, the U.S. Department of Agriculture, the World Health Organization, and others. The center's efforts include collecting, curating, and disseminating genomic data on bacterial resistance, which can help provide researchers with key data to support the development of promising antibiotic candidates.
Food and Drug Administration (FDA)	FDA conducts and funds regulatory science research to advance scientific knowledge related to antibiotic development.a For example, in February 2018, FDA solicited input from industry stakeholders on research ideas related to antibiotic clinical trial design, evaluating endpoints for use in antibiotic clinical trials, and other topics. In addition, according to FDA officials, FDA awarded a contract in fiscal year 2018 to develop animal models for use in preclinical research on infections caused by two types of resistant bacteria: *Acinetobacter baumannii* and *Pseudomonas aeruginosa*, and FDA is internally developing animal models for bacteriophage research.
	FDA hosts workshops related to the development of treatments for antibiotic resistance.
	According to FDA officials, FDA coordinates regularly with drug regulators in Europe, Canada, and Japan to discuss regulatory science issues about antibiotics that all three agencies face and strive to harmonize clinical trial requirements across jurisdictions where possible.

Table 5. (Continued)

Agency	Efforts
Department of Health and Human Services	
Centers for Disease Control and Prevention (CDC) and FDA	With input from FDA, CDC manages and operates the Antimicrobial Resistance Isolate Bank, which scientists can use when conducting antibiotic research and development. The bank contains isolates—pure samples of a bacteria—that can be used, for example, to study bacterial mechanisms of resistance or to test a drug's effectiveness. As of January 2018, the isolate bank shipped more than 2,000 isolate panels to researchers, including drug developers, diagnostic test developers, and other scientists. The Walter Reed Army Institute of Research also maintains a collection of antimicrobial resistance isolates collected from military hospitals.
Department of Defense (DOD)	
Defense Threat Reduction Agency	According to DOD officials, the agency has awarded 21 projects to study treatments for antibiotic-resistant infections, totaling approximately $178 million since 2012. In addition, the agency has funded 12 projects totaling nearly $50 million for other antibiotic resistance-related matters. Grantees include universities, the U.S. Army Medical Research Institute, and a national laboratory.
U.S. Army Medical Research and Materiel Command[b]	According to DOD officials, the command has funded 50 projects to study treatments for antibiotic-resistant infections since 2012, totaling $66.2 million. Funded projects study a range of product types, including traditional antibiotics, bacteriophages, peptides, and others. The command also funds 13 projects that study vaccines and other preventive products related to antibiotic-resistant infections.
Walter Reed Army Institute of Research	According to DOD officials, the institute supports 16 projects—five intramural and 11 extramural—totaling nearly $10 million since 2016. All of the projects are studying traditional antibiotic drug candidates, and all but one target gram negative bacteria. Two of the projects are conducted jointly, through an interagency agreement, with NIH institutes. In addition, the institute runs several other internal projects, including one to develop monoclonal antibodies, one to develop animal models to test potential antibiotic treatments, and one to conduct basic research on gram negative bacteria. The institute also conducts a bacteriophage project in partnership with a private company, which, as of January 2019, had completed phase 1 clinical trials.
Defense Advanced Research Projects Agency	Through its Pathogen Predators program, the agency studied living bacteria that prey upon pathogenic gram-negative bacteria. According to DOD officials, the agency awarded a total of $17.1 million in fiscal years 2014 through 2019 to four grantees.
Navy Medical Research Center	According to DOD officials, Navy researchers are investigating several types of nontraditional products, such as bacteriophages and probiotics, in collaboration with DOD partners, academic researchers, and industry partners.

Agency	Efforts
Department of Defense (DOD)	
	The bacteriophage research, which aims to develop treatments for *A. baumannii* and *Staphylococcus aureus* bacterial infections, is conducted jointly with the Walter Reed Army Institute of Research and a private company.
Department of Energy	
National Laboratories	According to Department of Energy officials, scientists conduct intramural research to understand the mechanisms bacteria develop to become resistant to antibiotics and how to overcome those mechanisms. For example, scientists at Lawrence Livermore National Laboratory have studied the mechanism by which microbes make compounds to become resistant to antibiotics and how enzymes in bacteria can be manipulated to allow antibiotics to pass through the bacterial cell wall. Scientists at the Sandia National Laboratory and a university have used computer models to understand how efflux pumps—which can remove antibiotics from the bacterial cell—work, and how to overcome them.

Source: GAO (summary), Departments of Health and Human Services, Defense, and Energy (data). | GAO-20-341.

[a]Regulatory science is defined by the FDA as the science of developing new tools, standards, and approaches to assess the safety, efficacy, quality, and performance of all FDA-regulated products. We previously reported that FDA lacked measurable goals to assess its progress in advancing regulatory science and recommended FDA develop and document measurable goals, including targets and time frames, and systematically track funding across its regulatory science priority areas. As of January 2020, FDA has not yet implemented these recommendations. See GAO, *Medical Product Oversight: FDA Needs More Strategic Planning to Guide Its Scientific Initiatives*, GAO-16-432 (Washington, D.C.: May 16, 2016).

[b]According to DOD officials, The U.S. Army Medical Research and Materiel Command was redesignated as the U.S. Army Medical Research and Development Command in June 2019.

Appendix IV: Additional Information on Federal Efforts to Promote Appropriate Antibiotic Use

This appendix contains more detailed information on federal efforts to promote the appropriate use of antibiotics in health care through antibiotic stewardship programs and activities, organized by agency. These examples do not comprise the full extent of agencies' efforts.

Table 6. Examples of federal efforts to promote the appropriate use of antibiotics

<table>
<tr><th>Agency</th><th>Efforts</th></tr>
<tr><td colspan="2">Department of Health and Human Services (HHS)</td></tr>
<tr><td>Centers for Disease Control and Prevention (CDC)</td><td>CDC published a series of guidance documents to promote the appropriate use of antibiotics in health care—called the Core Elements of Antibiotic Stewardship (Core Elements)—from 2014 to 2018.[a] CDC tailored the Core Elements to hospitals, nursing homes, outpatient settings, small and critical access hospitals,[b] and low- and middle-income countries with limited resources. Common elements in these guidance documents include (1) leadership commitment, (2) implementation of policies and interventions to improve antibiotic use, (3) tracking and reporting antibiotic use, and (4) education to providers on appropriate antibiotic use. While health care facilities and clinicians are not required to implement the Core Elements, experts we interviewed told us these guidance documents provided an important foundation for establishing antibiotic stewardship programs and conducting related activities.
Examples of antibiotic stewardship policies and interventions that health care providers can implement, depending on the setting, to support optimal antibiotic prescribing are:

• writing formal statements in support of improving antibiotic use;
• identifying individual leaders to be accountable for antibiotic stewardship activities;
• documenting antibiotic dose, duration, and indication in patients' medical records;
• developing and implementing facility-specific treatment recommendations, based on national guidelines and local susceptibilities;
• requiring prior authorization of certain antibiotics by an expert in antibiotic use, or working with a consultant pharmacist, before therapy is initiated;
• using delayed prescribing practices or watchful waiting, when appropriate;
• giving antibiotic "time-outs" 48 hours after initial treatment to assess the continuing need and choice of antibiotics after obtaining diagnostic information;
• automatically switching from intravenous to oral antibiotic therapy in appropriate situations, which improves patient safety by reducing intravenous access; and
• providing prospective audit and feedback, conducted by an expert in antibiotic use.</td></tr>
</table>

Agency	Efforts
Department of Health and Human Services (HHS)	
	CDC has expanded its collection of antibiotic use data across health care settings. Since we reported on antibiotic use data gaps in 2011,[c] CDC has obtained additional data regarding both inpatient and outpatient settings through its own surveillance systems as well as from other sources. In particular, CDC has focused its efforts to expand antibiotic use data collection from hospitals, where an estimated one in two patients receives an antibiotic for at least 1 day during an average hospital stay. CDC launched its Antimicrobial Use (AU) Option in 2011 as a voluntary, electronic reporting tool added on to the pre- existing National Healthcare Safety Network (NHSN). The AU Option allows the nation's 6,849 hospitals (as of January 1, 2020), including 13 Department of Defense (DOD) hospitals located on military bases outside the United States, that are already reporting to the NHSN to submit their antibiotic use data in a standardized format. CDC then aggregates such data to calculate national benchmarks and allows hospitals to compare their actual ("observed") antibiotic use against those benchmarks ("predicted" use). In addition, CDC has periodically conducted prevalence surveys through its Emerging Infections Program to gather data on health care-associated infections and antibiotic use in about 200 hospitals and 161 nursing homes in 10 states. With regard to outpatient settings, CDC has acquired, through a proprietary source, 8 years of pharmacy data on antibiotic prescriptions since 2011, which the agency is using to better characterize patterns in outpatient prescribing and to develop targeted interventions for high-prescribing areas. Through partnerships with other federal agencies, state and local health departments, and others, CDC has supported these efforts and published several analyses of antibiotic use—including state-level prescribing rates, the conditions for which antibiotics were prescribed, and the appropriateness of their use. For example, CDC has reported on geographic variations in antibiotic prescriptions in outpatient settings, by state, across the United States.
	To gauge progress toward one of the "significant outcomes" included in the *National Action Plan for Combating Antibiotic Resistance* (National Action Plan)—for all U.S. acute care hospitals to establish antibiotic stewardship programs by 2020—CDC tracks whether such hospitals have implemented antibiotic stewardship programs meeting each of the Core Elements. According to CDC, 85 percent of the nation's acute care hospitals reported having antibiotic stewardship programs in 2018 that met all seven of the Core Elements, compared with 41 percent in 2014.

Table 6. (Continued)

Agency	Efforts
Department of Health and Human Services (HHS)	
	In 2018, CDC launched a free, online training course for various types of clinicians—including physicians, dentists, pharmacists, physician assistants, and nurses—to inform them about appropriate antibiotic prescribing, how to overcome barriers, and strategies for communicating with patients. Clinicians can receive credit for partial completion (50 percent or more), or full completion, of this training as improvement activities under the Merit-based Incentive Payment System (MIPS) in 2019. In addition, clinicians who prescribe antibiotics, as well as other types of health care professionals, are eligible to receive up to 8 hours of continuing education credit for completing this training. As of late July 2019, nearly 16,000 individuals had registered for this training, according to CDC officials. Agency officials also told us that CDC has collaborated with medical schools, professional health care organizations, and others to develop and promote antibiotic stewardship training. For example, CDC collaborated with Wake Forest University's School of Medicine and the Association of American Medical Colleges to develop a model curriculum on antibiotic stewardship for medical students.[d]
	CDC supports research to identify, develop, and implement practices to stop the spread of resistance and to promote appropriate use of antibiotics in health care. CDC also supports research to fill gaps in knowledge related to aspects of antibiotic use and resistance that have public health impact. According to agency officials, CDC has provided approximately $110 million since 2016 to support this research through cooperative agreements and contracts.
	In 2017, CDC revised a national campaign to promote public awareness about appropriate antibiotic use. The campaign, called "Be Antibiotics Aware: Smart Use, Best Care," is aimed at both health care providers and the general public. This campaign refines the message from CDC's earlier campaign ("Get Smart: Know When Antibiotics Work") and has expanded to new target audiences, such as adult patients, physicians who work in hospitals, nurse practitioners, and physician assistants. To support the campaign, CDC has produced a number of materials and resources, including posters, brochures, and fact sheets for use and display in physicians' offices, pharmacies, and other health care and public settings. In addition, CDC has embedded stewardship messaging into other activities, including its "Get Ahead of Sepsis" campaign, which emphasizes the importance of early recognition and timely treatment of sepsis, reassessment of antibiotic therapy, and preventing infections that can lead to sepsis. CDC views these two campaigns as promoting the integration of antibiotic stewardship into sepsis management.

Agency	Efforts
Department of Health and Human Services (HHS)	
	CDC works with countries around the world to combat antibiotic resistance. For example, in conjunction with HHS's Office of Global Affairs, CDC launched the Antimicrobial Resistance Challenge at the United Nations General Assembly in September 2018 to catalyze global action against antibiotic resistance. A year later, CDC announced this challenge had resulted in nearly 350 commitments from government health officials, pharmaceutical and health insurance companies, and others from 33 countries to make formal commitments that further the progress against antimicrobial resistance, such as by improving appropriate antibiotic use.
Centers for Medicare & Medicaid Services (CMS)	In September 2019, CMS finalized new health and safety requirements for the nation's hospitals and critical access hospitals to implement antibiotic stewardship programs, as a condition of their participation in the Medicare and Medicaid programs, by March 30, 2020.[e] Under these requirements, hospitals and critical access hospitals are required, among other things, to implement these programs facility-wide (which includes emergency departments) and to adhere to nationally recognized guidelines for prescribing antibiotics. These requirements also apply to inpatient rehabilitation facilities (which are considered post- acute care facilities) and long-term acute care hospitals. In October 2016, CMS published a rule that established similar requirements for nursing homes as well as skilled nursing facilities—collectively known as long-term care facilities—to implement antibiotic stewardship programs.[f] Under this rule, long-term care facilities were required, in part, to have an antibiotic stewardship program in place by December 4, 2017. To complement the new requirements, CMS published new interpretive guidance on how to implement and survey for antibiotic stewardship in long-term care facilities, citing CDC's Core Elements.[g] CMS also updated its training webinars for surveyors—who monitor compliance with CMS's health and safety standards using on-site surveys—to include information on antibiotic use in nursing homes.
	CMS has developed incentives for eligible clinicians in any type of health care facility to improve antibiotic use and stewardship, as part of the agency's broader efforts to improve care for Medicare patients.[h] Through MIPS, launched in 2017, CMS offers hundreds of quality measures and nearly 100 "improvement activities" on a wide range of topics, including the appropriate use of antibiotics, that eligible clinicians can choose from and report their performance on to the agency. CMS then adjusts payments higher for clinicians who report data and achieve a performance-based, final score above a certain threshold—and penalizes clinicians who do not achieve that threshold with lower payments. In 2014, CMS added antibiotic stewardship to its efforts to provide better care at lower cost for Medicare beneficiaries in outpatient settings.

Table 6. (Continued)

Agency	Efforts
Department of Health and Human Services (HHS)	
	Specifically, CMS tasked its networks of health quality experts, clinicians, and others (called Quality Improvement Networks) with helping 7,629 outpatient facilities— including physician practices, emergency departments, and urgent care centers—implement CDC's Core Elements. Under this 5-year project, CMS is providing outreach, training, and technical assistance to providers and patients to encourage the expansion of antibiotic stewardship. CMS has provided training, technical assistance, and other learning opportunities to more than 4,000 hospitals, 2,400 nursing homes, and 7,600 outpatient settings on best practices for antibiotic stewardship and guidance on *C. difficile* prevention. In addition, CMS and CDC have developed and launched free, online training to help nursing homes implement antibiotic stewardship and prevent and manage *C. difficile* infections.
Agency for Healthcare Research and Quality (AHRQ)	AHRQ has increased its support since 2015 for research to develop improved methods to combat antibiotic resistance and promote antibiotic stewardship, including through grants for research that will total more than $57 million, according to AHRQ officials. This research includes studies on the role of diagnostic tools, such as blood cultures, in improving antibiotic use and reducing antibiotic resistance. AHRQ has also published numerous research studies on antibiotic or antimicrobial stewardship that the agency funded or authored.
	Through a 5-year nationwide project, the AHRQ Safety Program for Improving Antibiotic Use has provided technical assistance and CDC's guidance to hospitals, long-term care facilities, and physicians' offices to promote implementation of antibiotic stewardship activities and to help clinicians select optimal antibiotic treatment regimens. In December 2018, AHRQ completed implementation of this guidance in more than 400 hospitals, which included six DOD facilities and 79 critical access hospitals, according to AHRQ officials. AHRQ officials also told us that preliminary data suggest that antibiotic use was reduced in this cohort, and an educational toolkit based on this cohort's activities will be released in fiscal year 2020. The project has also recruited more than 450 nursing homes for a 1-year cohort planned to be completed in November 2019, and AHRQ planned to expand to an additional 250 to 500 ambulatory care settings in December 2019. AHRQ also plans to compile tools, resources, and lessons learned to promote implementation of antibiotic stewardship in these health care settings.
	AHRQ has increased dissemination of its Nursing Home Antimicrobial Stewardship Guide, which provides toolkits to help nursing home staff create an antibiotic stewardship program, determine whether to treat with antibiotics, choose the right antibiotic, and engage residents and families.[i]

Agency	Efforts
Department of Health and Human Services (HHS)	
Health Resources and Service Administration (HRSA)	In November 2017, HRSA's Federal Office of Rural Health Policy added a requirement for critical access hospitals participating in the Medicare Beneficiary Quality Improvement Project to implement antibiotic stewardship programs by the end of fiscal year 2021 in order to be eligible for the Medicare Rural Hospital Flexibility grant program.[j] According to HRSA officials, this requirement will apply to 99 percent of the nation's more than 1,300 critical access hospitals.
Office of Global Affairs	To promote the appropriate use of antibiotics internationally, the Office of Global Affairs has collaborated with other countries and co-chairs the Transatlantic Taskforce on Antimicrobial Resistance (TATFAR) program. With respect to antibiotic stewardship, TATFAR working groups have (1) developed a common structure and indicators for antibiotic stewardship programs; (2) reviewed antibiotic use reduction goals; (3) published a resource for other countries seeking to improve their antibiotic use; (4) aligned campaigns promoting appropriate antibiotic use by collaborating with partners such as the World Health Organization in supporting World Antibiotic Awareness Week; (5) compiled resources for how to assess appropriate outpatient use in TATFAR countries; and (6) published studies on health care-associated infections and antibiotic use across health care facilities.
Department of Defense (DOD)	
Defense Health Agency	DOD published a policy, effective October 2017, requiring the establishment of antibiotic stewardship programs within its military medical treatment facilities and issued guidance for implementation in October 2018. These facilities include hospitals located at military installations across the United States and abroad, plus ambulatory care clinics and dental clinics, and provide health care for active-duty service members, their dependents, and other eligible beneficiaries. The policy calls for these facilities to establish antibiotic stewardship programs including the following components, at a minimum: (1) leadership commitment by each facility; (2) accountability; (3) pharmacy expertise, including antibiotic prescribing and use evaluation; (4) the creation of antibiograms—an aggregate profile of antibiotic susceptibility for a given facility, locality, region, or nation; (5) the identification and implementation of at least one action for change that would demonstrate commitment to the program; (6) training for clinicians regarding antibiotic resistance and prescribing practices; and (7) facility-specific treatment recommendations in accordance with local antibiograms. DOD officials told us that all of these facilities (both inpatient and outpatient) were in different stages of implementing the antibiotic stewardship policy, as of August 2019. In addition, as part of these requirements, DOD hospitals are to report their antibiotic use data to CDC through the NHSN

Table 6. (Continued)

Agency	Efforts
Department of Veterans Affairs (VA)	
	AU Option. DOD officials told us that 44 of DOD's 48 hospitals have reported their antibiotic use data to CDC, as of September 30, 2019.
	According to department officials, DOD offers antibiotic stewardship training to its health care facilities through webinars, workshops, and briefings.
Veterans Health Administration	VA's efforts to implement antibiotic stewardship programs in its health care facilities, which provide both inpatient and outpatient services to veterans, pre-date the National Action Plan and CDC's Core Elements. In January 2019, VA updated its 2014 policy directive for the implementation and maintenance of antibiotic stewardship programs that, in part, requires each of its health care facilities to (1) develop a written policy, (2) perform an annual evaluation of stewardship activities, (3) ensure that adequate staff and resources are in place, and (4) identify medical and pharmacy personnel as stewardship "champions." The updated policy directive also requires all VA facilities with 30 or more acute care beds to report their antibiotic use data to CDC's NHSN AU Option by January 30, 2020. According to department officials, VA has successfully implemented antibiotic stewardship programs in all of its health care facilities, and 113, or more than 97 percent of, 116 eligible VA hospitals with 30 or more acute care beds were reporting their antibiotic use data to CDC, as of January 1, 2020.
	VA offers antibiotic stewardship training to its health care facilities through webinars on an ongoing basis.

Source: GAO analysis of information from HHS, DOD, and VA. | GAO-20-341

[a]Department of Health and Human Services, Centers for Disease Control and Prevention, *The Core Elements of Hospital Antibiotic Stewardship Programs* (Atlanta, Ga.: 2014) and *The Core Elements of Hospital Antibiotic Stewardship Programs: 2019* (Atlanta, Ga.: 2019), *The Core Elements of Antibiotic Stewardship for Nursing Homes* (Atlanta, Ga.: 2015), *The Core Elements of Outpatient Antibiotic Stewardship* (Atlanta, Ga.: 2016), *Implementation of Antibiotic Stewardship Core Elements at Small and Critical Access Hospitals* (Atlanta, Ga.: 2017), and *The Core Elements of Human Antibiotic Stewardship Programs in Resource-Limited Settings: National and Hospital Levels* (Atlanta, Ga.: 2018).

The Joint Commission—a nonprofit organization that accredits and certifies more than 22,000 health care organizations and programs in the United States—added new accreditation requirements, effective in 2017, for hospitals and nursing homes to implement antibiotic stewardship programs consistent with CDC's Core Elements. On the outpatient side, The Joint Commission finalized new requirements in June 2019 for ambulatory health care organizations that it accredits to implement antimicrobial stewardship activities; these requirements go into effect in 2020. Another organization, DNV GL Healthcare USA, Inc., also issued an antibiotic stewardship standard in 2018, based on CDC's Core Elements, for the hospitals it accredits and certifies.

[b]"Critical access hospital" is a designation given to eligible rural hospitals by states and certified by CMS. 42 U.S.C. § 1395i-4(c)(2)(B), 42 C.F.R. § 485.606(b). Congress created the critical access hospital designation through the Balanced Budget Act of 1997 in response to a string of rural hospital closures in the 1980s and early 1990s. Among other criteria, eligible hospitals must have 25 or fewer acute-care inpatient beds and be located more than 35 miles from another hospital, although exceptions may apply. See Balanced Budget Act of 1997, Pub. L. No. 105-33, § 4201, 111 Stat. 251, 369-71 (codified in pertinent part at 42 U.S.C. §§ 1395f(l), 1395i-4, 1395m(g), (l)(8)). As of October 2019, there were 1,349 critical access hospitals in the United States.

[c]GAO, *Antibiotic Resistance: Data Gaps Will Remain Despite HHS Taking Steps to Improve Monitoring*, GAO-11-406 (Washington, D.C.: June 1, 2011).

[d]Wake Forest School of Medicine, "Antimicrobial Stewardship Curriculum," accessed December 18, 2019, https://school.wakehealth.edu/Departments/Internal-Medicine/Infectious- Diseases/ CAUSE/Antimicrobial-Stewardship-Curriculum.

[e]Department of Health and Human Services, Centers for Medicare & Medicaid Services, *Medicare and Medicaid Programs; Regulatory Provisions to Promote Program Efficiency, Transparency, and Burden Reduction; Fire Safety Requirements for Certain Dialysis Facilities; Hospital and Critical Access Hospital (CAH) Changes to Promote Innovation, Flexibility, and Improvement in Patient Care*, 84 Fed. Reg. 51732 (Sept. 30, 2019) (pertinent provisions to be codified at 42 C.F.R. §§ 482.42 (d), 485.640).

[f]Department of Health and Human Services, Centers for Medicare & Medicaid Services, *Medicare and Medicaid Programs; Reform of Requirements for Long-Term Care Facilities*, 81 Fed. Reg. 68688 (Oct. 4, 2016) (pertinent provision codified at 42 C.F.R. § 483.80(a)(3) (2018)).

[g]Centers for Medicare & Medicaid Services, *State Operations Manual Appendix PP – Guidance to Surveyors for Long-Term Care Facilities* (Nov. 22, 2017).

[h]The Medicare Access and CHIP Reauthorization Act of 2015 required CMS to implement an incentive program for clinicians participating in Medicare to receive higher payments based on their performance. Medicare Access and CHIP Reauthorization Act of 2015, Pub. L. No. 114-10, 129 Stat. 87, 92 (Apr. 16, 2015), codified at 42 U.S.C. § 1395w-4. In 2019, MIPS-eligible clinician types include physicians (doctors of medicine, which includes many specialties; doctors of dental surgery or dental medicine; and doctors of osteopathy), physician assistants, nurse practitioners, and others. In addition, to be eligible for MIPS in 2019, the clinician must (1) bill more than $90,000 in allowed charges for professional services covered under the Medicare Physician Fee Schedule; (2) provide more than 200 covered professional services; and (3) furnish covered professional services to more than 200 Medicare beneficiaries. Eligible clinicians may participate in MIPS as individuals or as part of a group that includes one or more of the eligible clinician types.

[i]Agency for Healthcare Research and Quality, "Nursing Home Antimicrobial Stewardship Guide," accessed December 17, 2019, https://www.ahrq.gov/nhguide/index.html.

[j]HRSA is the primary federal agency for improving health care to people who are geographically isolated or economically or medically vulnerable, according to the agency's website; tens of millions of Americans receive quality, affordable health care and other services through its 90-plus programs and more than 3,000 grantees.

Appendix V: Comments from the Department of Health and Human Services

DEPARTMENT OF HEALTH & HUMAN SERVICES OFFICE OF THE SECRETARY

Assistant Secretary for Legislation
Washington, DC 20201

MAR 0 6 2020

Mary Denigan-Macauley
Director, Health Care
U.S. Government Accountability Office
441 G Street NW
Washington, DC 20548

Dear Ms. Denigan-Macauley:

Attached are comments on the U.S. Government Accountability Office's (GAO) report entitled, *"Antibiotic Resistance: Additional Federal Actions Needed to Better Determine Magnitude and Reduce Impact"* (GAO-20-341).

The Department appreciates the opportunity to review this report prior to publication.

Sincerely,

Sarah Arbes
Acting Assistant Secretary for Legislation

GENERAL COMMENTS FROM THE DEPARTMENT OF HEALTH & HUMAN SERVICES ON THE GOVERNMENT ACCOUNTABILITY OFFICE'S DRAFT REPORT ENTITLED — ANTIBIOTIC RESISTANCE: ADDITIONAL FEDERAL ACTIONS NEEDED TO BETTER DETERMINE MAGNITUDE AND REDUCE IMPACT (GAO-20-341)

The U.S. Department of Health & Human Services (HHS) appreciates the opportunity from the Government Accountability Office (GAO) to review and comment on this draft report.

National Action Plan for Combating Antibiotic Resistant Bacteria (CARB) (NAP) Not Fully Addressed

Antibiotic resistance is a large scale and complex topic for which the United States Government has developed the National Action Plan for Combating Antibiotic Resistant Bacteria (CARB)(NAP) to address. HHS notes that the depth and breadth of current activities of agencies and departments under this comprehensive national plan are not fully reflected in this report, and actions to address these recommendations and other priorities would be necessary to fully achieve CARB goals.

There are two fundamental and important topics included in the USG CARB NAP that are not included in the GAO report that HHS finds important to highlight. Infection prevention and control is one of the foundational activities described under Goal 1 of the CARB NAP, and it is fundamentally needed to address the emergence of the spread of resistant pathogens in healthcare, food and the community. GAO's report provides limited information to highlight the importance of this topic, and HHS notes the importance of including the need for proper infection prevention and control in addressing the spread of antibiotic resistant infections. As highlighted in CDC's recent 2019 Antibiotic Resistance Threats Report, prevention efforts such as these have reduced deaths from antibiotic-resistant infections by 18 percent overall and by nearly 30 percent in hospitals since 2012. HHS views the discussion of infection prevention and control as on par with the other elements mentioned in the report (such as surveillance, stewardship, diagnostics, and drug development), and as essential to give a complete picture of the challenges, needs, and progress to date in this critically important area.

In addition, while HHS understands that the GAO's scope for this report was limited to antibacterial resistance, we note that the report excludes resistant fungal infections from its discussion and analysis. The pathogens covered under the US CARB National Action Plan are those identified by CDC as threats in its AR Threats Report. This includes bacterial and fungal pathogens, and importantly CDC's most recent Threats Report identified *Candida auris* as an urgent threat – the highest threat category – in the United States. The needs and actions to be taken to address fungal resistance are largely the same as those needed to address bacterial resistance and acknowledging the threat of fungal resistance is vital to fully reflecting the scope of the public health challenges that we face domestically and globally. . When individuals become sick with bacterial or fungal infections, they want to be confident that an effective treatment exists or better yet to have effective prevention measures that ensure they do not get an infection in the first place.

The development of the CARB National Strategy and the CARB NAP brought together departments and agencies spanning the U.S. government to comprehensively articulate national goals, activities, and milestones and report on their progress regularly and transparently. This collaboration has built and continues to build programs and national infrastructure that now lead the world on many fronts and has transformed how the country responds to the threats of antibiotic resistance in every key area. While these activities are far from complete and in fact many efforts need to be scaled up considerably, CARB work has placed the U.S. government in a position where we have the potential -- through ongoing activities and resources -- to get ahead of one of the world's most challenging public health threats. The forthcoming update to the CARB NAP will build on the success of the first Action Plan and push the U.S. to make further progress domestically and globally.

Antimicrobial Susceptibility Test Devices

HHS recognizes that this report accurately discusses many of the historical challenges associated with updating antimicrobial susceptibility testing (AST) devices to reflect the most current susceptibility test interpretive criteria (STIC), also known as "breakpoints," for systemic antibacterial or antifungal drugs.

HHS commends the efforts of Congress to address these challenges through Section 511A of the Federal Food, Drug, and Cosmetic Act (the FD&C Act) (21 U.S.C. 360a-2), as added by section 3044 of the 21st Century Cures Act (Cures Act) (Pub. L. 114-255), signed into law on December 13, 2016. This provision clarifies FDA's authority to identify and efficiently update susceptibility test interpretive criteria, including through the recognition by FDA of standards established by standard development organizations (SDOs). It also clarifies that manufacturers of AST devices may rely upon FDA-recognized breakpoints to support premarket authorization of their devices, provided they meet certain conditions, which provides for a more streamlined process for incorporating up-to-date information into such devices.

Since passage of the Cures Act, FDA has worked to implement section 511A of FD&C Act, and engaged in other efforts to promote the availability and updating of AST devices. In accordance with section 511A, FDA established in December 2017, within 1 year of enactment of the Cures Act, an Interpretative Criteria Website[1] (STIC website) that contains a list of FDA-recognized susceptibility test interpretive criteria standards (also referred to as breakpoints or STIC), as well as other susceptibility test interpretive criteria identified by FDA. This website identifies when FDA does or does not recognize breakpoints established by an SDO and lists breakpoints identified by FDA outside the SDO process. To ensure that the most up-to-date STIC information is available, holders of approved drug applications were required to remove STIC and related information from the labeling of approved systemic antibacterial and antifungal drugs and replace it with a reference to the STIC website.[2] AST device developers may rely upon the breakpoints listed on the FDA STIC website to support premarket submission of their devices. Further, in accordance with section 511A(c)(1) of the FD&C Act, at least every 6 months, FDA publishes on the website a notice recognizing new or updated susceptibility test interpretive criteria standards, or parts of standards; withdrawing recognition of susceptibility test interpretive criteria standards, or parts of standards; and making any other necessary updates to the lists published on the website. This makes up-to-date information about antimicrobial

---

[1] See https://www.fda.gov/STIC.

[2] https://www.fda.gov/media/109839/download. See section 511A(d) of the FD&C Act.

susceptibility easily accessible by device manufacturers that are considering any changes to their devices and labeling.

In further efforts to promote the availability of up-to-date AST devices, FDA published draft guidance in early 2016 and, ultimately, finalized that guidance in February 2019, on use of a coordinated development process for new antimicrobial drugs and new or existing AST devices. Implementation of the final guidance, *Coordinated Development of Antimicrobial Drugs and Antimicrobial Susceptibility Test Devices,*[3] has drastically improved the time for AST device availability following new drug approval, with devices becoming available within weeks of drug approval in most cases.

Additionally, FDA has worked closely with AST device manufacturers and other stakeholders to understand the challenges associated with incorporating breakpoint updates applicable to their devices in a timely fashion. Through these discussions, FDA identified the opportunity to encourage the development of breakpoint change protocols for new and existing AST devices. AST device manufacturers can submit a breakpoint change protocol in their premarket submissions, outlining how they will validate breakpoint updates incorporated into their device when such updates are recognized by FDA and posted to the STIC website. Following change protocol submission and concurrence by FDA, AST device manufacturers may implement breakpoint changes to their devices and labeling to reflect the most current information without the need for a 510(k) submission, if certain conditions are met. This may be done when the inclusion and performance criteria outlined in the breakpoint change protocol are confirmed by the manufacturer. FDA reviewed and concurred with the first breakpoint change protocol in 2018, and FDA has continued to encourage manufacturers to use this approach. FDA expects that this approach will expedite updating of devices and assist in monitoring the status of these devices following FDA recognition of new or updated breakpoints.

Further, FDA intends to continue regular engagement with AST device manufacturers and related stakeholders to continue to improve these processes. While stakeholders can sign up for email updates when the STIC website is updated, FDA also intends to individually reach out to AST device manufacturers when breakpoint updates are made that could potentially affect their cleared devices. FDA hopes this additional outreach and coordination will lead to more rapid updates of AST devices.

While there is still much to be done, HHS is proud of how far we have come as a country and continues to look forward to how we can further address the ongoing threats of antibiotic resistance.

---

[3] https://www.fda.gov/regulatory-information/search-fda-guidance-documents/coordinated-development-antimicrobial-drugs-and-antimicrobial-susceptibility-test-devices

Recommendation 1
The Director of CDC should take steps to determine participation rates and distribution needed in the AR Option of the National Healthcare Safety Network for conducting regional and national assessments of antibiotic resistance of public health importance.

HHS Response
HHS concurs with GAO's recommendation.
CDC recognizes the concerns of GAO and the Joint Public Health Informatics Task Force regarding the breadth of data currently available in the National Healthcare Safety Network (NHSN) Antimicrobial Use and Resistance (AUR) Module's AR Option. NHSN's AUR module helps healthcare facilities to track antimicrobial resistance patterns for more than 20 different organisms. As there is currently no national requirement for facilities to report in the AR option, CDC is actively working with public health partners, including other federal agencies, health departments, health information technology vendors, and healthcare systems to promote the voluntary use of the AR option.

While NHSN has experienced vast expansion over time, appropriated resources have stayed largely stable with only modest increases. Since its creation in 2005, NHSN has grown from 300 hospitals reporting to more than 22,000 in 2020, an increase of approximately 7,300 percent. Over this same time, appropriated resources for NHSN have only increased by 40 percent.

CDC also provides technical support to states that may be considering a state or local mandate to require AR/ AU reporting. To date, two states have implemented a local mandate that requires AR/AU reporting, and one state is considering implementation of a mandate.

With the current level of reporting, CDC is developing pilot programs to assess AR option data and other data sources for determining the landscape for regional and national extended-spectrum beta-lactamases (ESBLs). The goal of this analysis will be to improve our understanding of community transmission and amplification in healthcare settings in order to have greater impact on preventing ESBLs. This work is still in the early development phase; however, CDC believes it will be a valuable initial analysis of the use of AR option data beyond the hospital facility-level.

Recommendation 2
The Director of CDC should ensure that CDC's evaluation of its surveillance system for antibiotic-resistant gonorrhea include measures of its representativeness, such as comparison of the trends in the sample population with those in the overall U.S. population, using specially designed studies if needed.

HHS Response

HHS concurs with this recommendation.

CDC concurs with the recommendation to assess the representativeness of GISP. It should be noted, however, that GISP, as a sentinel surveillance system, was not designed to be representative of the U.S. population with gonorrhea (GC) but, rather to identify new trends in resistance before they become generalized across the entire population with GC. Historically, CDC has successfully used this approach to anticipate emerging trends and inform changes in the CDC's STD Treatment Guidelines to ensure patients are treated with the most effective antibiotic available.

Through eGISP and SURRG, CDC has taken steps to examine the representativeness of GISP as mentioned in the report. By testing women with gonorrhea, extragenital sites of men with gonorrhea, and men and women from non-sexually transmitted disease clinic sites, results can be compared with those from GISP. It is currently necessary to obtain cultured isolates for these laboratory analyses. This is more difficult in women, and from extra-genital sites in men in whom infections are usually asymptomatic, than it is in symptomatic men included in GISP. To overcome this, CDC is currently working to develop laboratory methods to examine genetic markers for GC resistance in swab specimens and urine without the need for cultured isolates. Although this would not replace culture-based surveillance, this would allow CDC to obtain a larger number of samples from non-GISP populations and to compare the genetic markers in those populations with ones from GISP patients. Since culture is now rarely used outside of CDC sponsored projects, this approach could provide samples that are more representative of the overall U.S. population infected with GC. However, additional resources are needed for developing and validating these methods and applying them to different populations.

In FY 2016, CDC requested $264 million annually to support full implementation of all of its activities under the CARB Action Plan including the expansion of GC testing nationwide. Ultimately, Congress supported funding for the overall CARB budget initiative of $160 million in FY 16 and has provided small increases since then. With available resources, CDC has supported an expansion of GC testing under the ARLN compared to pre-CARB levels, however, additional work to further refine and improve on the collection of AR data for GC cannot be completed with existing resources.

Recommendation 3

The Director of CDC should provide information on uncertainties for antibiotic resistance estimates in its consolidated Threats Reports, including standard errors or confidence intervals, as appropriate.

HHS Response
HHS generally concurs with GAO's recommendation.
CDC believes that information on uncertainties including standard errors or confidence intervals should be accessible within the public domain and is committed to doing so. Given that many of the new estimates provided in the 2019 AR Threats Report were based on new methodologies that aggregated data across three electronic health databases, CDC felt that it was critical to have the analysis peer reviewed by external experts and published in the scientific literature for wider reference and dissemination. CDC feels that publishing these data in a way that it is most helpful for the scientific community is critical, and multiple publications are pending. Once published in the scientific literature, CDC will link these publications back to online resources of the 2019 AR Threats Report for awareness and understanding of the connection between the report and the contributing peer reviewed material.

Recommendation 4
The Director of CDC should develop a plan for timely, consolidated reports of antibiotic resistance in priority pathogens at regular intervals.

HHS Response
HHS concurs with GAO's recommendation. The GAO report appears to imply that the AR Threats Report is the only report released by the agency that provides information on priority pathogens at regular intervals. To clarify, CDC already provides timely, consolidated updates of antibiotic resistance in priority pathogens at regular intervals through the following:

- NARMS Now (Updated Daily): CDC updates the latest approved resistance data for enteric pathogens nightly.
- AR & Patient Safety Portal (Updated Bi-annually): CDC updates resistance data for healthcare infections twice annually in the spring and fall.
- Active Bacterial Core surveillance (ABC) Bact Facts interactive (Updated annually): CDC includes resistance data for Streptococcus in annual updates.
- STD Surveillance Report & GISP (Updated Annually): Antibiotic resistant data for gonorrhea is released annually in the STD Surveillance Report and GISP Profiles with site-specific data.
- Reported Tuberculosis in the United States (annual surveillance report): Includes data on drug-resistant TB.

For all of these pathogens, other publications highlighting antibiotic resistance (MMWR, peer review publications) are issued ad hoc depending on new findings. Finally, CDC has plans to update its enterprise-wide AR Threats Report every three years.

Recommendation 5
The Secretary of HHS should identify leadership and clarify roles and responsibilities among HHS agencies to assess the clinical outcomes of diagnostic testing for identifying antibiotic-resistant bacteria.

HHS Response

HHS concurs with GAO's recommendation.

As part of the development of the next National Action Plan for CARB (2020-2025), CARB Task Force leadership will work with relevant HHS agencies to clarify roles and responsibilities for supporting research on clinical outcomes related to diagnostic tests. These discussions will include identifying leadership for this work, if such a strategy would be appropriate. The next Plan will maintain the goal of advancing the development and use of diagnostic tests (Goal 3).

Recommendation 6

The Commissioner of FDA should direct the Center for Devices and Radiological Health to conduct additional monitoring and evaluation of the status of commercial tests that rely on breakpoints, on a regular basis, to determine whether test manufacturers are updating breakpoints, seeking additional resources as needed.

HHS Response

HHS concurs with GAO's recommendation.

FDA concurs with GAO's recommendation to conduct additional monitoring and evaluation of AST devices to determine whether device manufacturers are making updates, where appropriate, when FDA identifies or recognizes new breakpoints.

FDA agrees with GAO that it is important that AST devices reflect the most current information as one aspect in improving patient care and antimicrobial stewardship, and combating antimicrobial resistance. Although FDA agrees that in the past there have been significant challenges associated with updating AST devices to reflect the most current breakpoints, FDA believes it is important to understand that many of the concerns raised in the report have been resolved by recent FDA actions and the passage and implementation of section 511A of the FD&C Act. As described above, FDA has taken major steps in the implementation of the new statutory provision, and has engaged in other activities to promote the availability and updating of AST devices.

Recommendation 7

The Secretary of HHS should develop a strategic framework to further incentivize the development of new treatments for antibiotic-resistant infections, including through the use of postmark financial incentives, and, if appropriate, make recommendations to Congress for necessary authority.

HHS Response

HHS non-concurs with GAO's recommendation because it is still unclear whether post-market financial incentives should necessarily be part of HHS's forthcoming strategic framework to further incentivize the development of new treatments to combat antibiotic resistance.

HHS agrees with GAO that additional incentives are needed to address the limited pipeline for novel and innovative treatments to combat antibiotic resistance. As noted by GAO, HHS convened a workgroup in March 2019 to analyze existing incentives, proposals for new incentives as raised by non-governmental, industry, and international groups, and other options to address this need. This analysis has included consideration of the current and future burden of AMR on both public health and the economy, the dynamics of drug development that specifically impact relevant antibacterial and related products, and the statutory, regulatory, and budgetary feasibility of any incentive strategy. Based on this work, the group is developing a strategic framework that includes specific proposals to address a variety of scientific and market challenges facing antibiotic product developers. ASPE and HHS are still conducting analyses to understand whether post-market financial incentives should be included in this framework as a solution among a range of potential solutions.

Recommendation 8

The Secretary of HHS should direct the CARB Task Force to include in its annual updates to the President plans for addressing any barriers preventing full implementation of the National Action Plan and, as appropriate, make recommendations for new or modified actions. Specifically, the CARB Task Force should identify plans to address barriers, such as those related to expanding (1) a CDC program designed to strengthen the U.S. response to resistant gonorrhea, (2) antibiotic stewardship programs across health care settings, and (3) antibiotic use data collection across health care settings, to the extent feasible

HHS Response

HHS concurs with GAO's recommendation.

Beginning in 2020 with the Annual Report for Year 5 of the first CARB National Action Plan, and continuing thereafter in the annual reports for the next iteration of the plan, HHS will include discussion of any barriers preventing full implementation of these activities, including any barriers that GAO has included in the recommendation where appropriate.

Beginning in 2020 with the Annual Report for Year 5 of the first CARB National Action Plan, and continuing thereafter in the annual reports for the next iteration of the plan, the CARB Task Force will include discussion of any barriers preventing full implementation of these activities, including the barriers that GAO has included in this recommendation where appropriate

In: A Closer Look at Antibiotic Resistance ISBN: 978-1-53619-277-3
Editor: Victor Tsunoda

Chapter 2

COMBATING SUPERBUGS: U.S. PUBLIC HEALTH RESPONSES TO ANTIBIOTIC RESISTANCE*

Committee on Energy and Commerce

Tuesday, June 14, 2016
House of Representatives,
Subcommittee on Oversight and Investigations,
Committee on Energy and Commerce,
Washington, DC.

The subcommittee met, pursuant to call, at 10:00 a.m., in room 2322, Rayburn House Office Building, Hon. Tim Murphy (chairman of the subcommittee) presiding.

---

* This is an edited, reformatted and augmented version of Hearing before the Subcommittee on Oversight and Investigations, of the Committee on Energy and Commerce, House of Representatives, One Hundred Fourteenth Congress, Second Session, Serial No. 114–153, dated June 14, 2016.

Present: Representatives Murphy, McKinley, Burgess, Black-burn, Bucshon, Flores, Brooks, Mullin, Hudson, Collins, Upton (ex officio), DeGette, Schakowsky, Castor, Tonko, Clarke, Kennedy, Green, and Pallone (ex officio).

Staff Present: Gary Andres, Staff Director; Emily Felder, Counsel, Oversight and Investigations; Jay Gulshen, Staff Assistant; Brittany Havens, Oversight Associate, Oversight and Investigations; Charles Ingebretson, Chief Counsel, Oversight and Investigations; Alan Slobodin, Deputy Chief Counsel, Oversight; Dylan Vorbach, Deputy Press Secretary; Jeff Carroll, Minority Staff Director; Tiffany Guarascio, Minority Deputy Staff Director and Chief Health Advisor; Chris Knauer, Minority Oversight Staff Director; Una Lee, Minority Chief Oversight Counsel; Elizabeth Letter, Minority Professional Staff Member; and Andrew Souvall, Minority Director of Communications, Outreach and Member Services.

OPENING STATEMENT OF HON. TIM MURPHY, A REPRESENTATIVE IN CONGRESS FROM THE COMMONWEALTH OF PENNSYLVANIA

Mr. MURPHY. Good morning, meeting here of the Energy and Commerce Subcommittee on Oversight and Investigations. I do want to announce to members this hearing is here and not in Room 2123. I just want you to know. I was there, you were not. It was a very powerful, moving——

Mr. UPTON. There was a very long line of folks to get in.

Mr. MURPHY. There was, so I figured it must be this hearing, but my apologies. It happens.

So this is a hearing on U.S. public health response to antibiotic resistance. The subcommittee convenes this hearing today to examine public health responses to the challenge of antibiotic-resistant superbugs. One of the world's most pressing health problems is the emergence of bacterial infections that are resistant to antibiotics. And according to the

Centers for Disease Control and Prevention, each year, 2 million Americans become sick with antibiotic-resistant infections, and of that, about 23,000 die. Globally, some institutions estimate up to 700,000 die each year from antibiotic-resistant infections, and without action, the researchers estimate 10 million people will die per year by 2050 from drug-resistant infections.

The World Health Organization has declared that humanity is on the precipice of a post-antibiotic era where common infections may once again be lethal because bacteria have become resistant to the antibiotics existing to treat them.

The antibiotic-resistance threat just got greater. Last month, a woman in my home State of Pennsylvania was diagnosed with an E. coli infection that had a rare gene called MCR–1, a new kind of superbug never before seen in the United States. Medical professionals were alarmed for two reasons. One, this new superbug is resistant to colistin, an antibiotic of last resort, which is used when no other antibiotics can fight the infection. And two, this MCR–1 gene can move from one bacteria to another. Eventually, MCR–1 could emerge with another superbug that is resistant to all antibiotics except for colistin and form an unstoppable superbug.

In response to the discovery of the MCR–1 gene, CDC Director Dr. Tom Frieden commented that the medicine cabinet is empty for some patients. It is the end of the road unless we act urgently. If this threat is not stopped, minor infections may become life-threatening and treatment for diseases such as cancer, diabetes, or routine surgeries will be at risk.

Fortunately, the end of the road is not here yet, and Congress and Federal agencies are working diligently to counter the effects of antibiotic resistance. These efforts confront the two main contributors to the spread of antibiotic resistance: the overprescription of antibiotics and the lack of new antibiotic development.

Since the discovery of penicillin in the early 20th century, almost every type of bacteria has become less responsive to antibiotics. As soon as an antibiotic goes into wide use among the general public, bacteria evolve to become resistant.

A study published last month in the Journal of the American Medical Association, known as JAMA, found that nearly a third of antibiotics prescribed in doctors' offices, emergency rooms and hospital-based clinics in the United States are not needed. This amounts to nearly 47 million unnecessary prescriptions given out each year. That is 47 million unnecessary prescriptions each year. And the number in this report most likely undercount the use of antibiotics because the data did not include urgent care clinics, retail pharmacies, dentist offices, and prescriptions given over the phone by nurse practitioners and physician assistants.

To combat antibiotic overuse, the CDC has partnered with the FDA to advocate for antibiotic stewardship programs in health care facilities throughout the United States. The CDC has issued guidelines about how hospitals can minimize inappropriate or excessive use of antibiotics, which could help reduce antibiotic overprescription. Reducing the inappropriate use of antibiotics will help, but it can only slow the spread of antibiotic-resistant bacteria, new antibiotics, and alternative therapies must be developed.

Despite the need for antibiotic development, as of March 2016 there were only 37 new antibiotics in development, and just 13 were in phase 3 clinical trials. These drugs would potentially address many resistant bacteria, but they are not enough. To combat this, in February of this year the Biomedical Advanced Research and Development Authority, known as BARDA, has collaborated with NIH to establish a biopharmaceutical accelerator that will support research and development to incentivize antibacterial drug development. This accelerator will, one, fund development of antibacterial products; and two, quickly move successful drug candidates through early development; three, provide business and drug development guidance; and four, decrease barriers to research and development of antibiotics.

The CDC, FDA, NIH, and BARDA have and continue to make significant and ongoing contributions to implement the National Action Plan for Combating Antibiotic-Resistant Bacteria released last year, which outlines steps to implement a national strategy to combat antibiotic resistance. Additionally, Congress has increased funding for these initiatives by 57 percent over last fiscal year for a total of more than $375 million.

Despite these promising developments, we are facing a public health challenge, and we need to ensure that the Federal Government is taking the appropriate action to protect the American public.

So I want to thank the witnesses for appearing here today before the subcommittee. I look forward to hearing all of your testimony on this very, very important issue.

[The prepared statement of Mr. Murphy follows:]

PREPARED STATEMENT OF HON. TIM MURPHY

The subcommittee convenes this hearing today to examine public health responses to the challenge of antibiotic resistant "superbugs."

One of the world's most pressing health problems is the emergence of bacterial infections that are resistant to antibiotics.

According to the Centers for Disease Control and Prevention, each year 2 million Americans become sick every year with antibiotic-resistant infections, and of that about 23,000 die.

Globally, some institutions estimate up to 700,000 die each year from antibiotic resistant infections. Without action, the researchers estimate 10 million people will die per year by 2050 from drug resistant infections.

The World Health Organization has declared that humanity is on the precipice of a "post-antibiotic era," where common infections may once again be lethal because bacteria have become resistant to the antibiotics existing to treat them.

The antibiotic-resistance threat just got greater. Last month, a woman in my home state of Pennsylvania was diagnosed with an E. coli infection that had a rare gene called MCR–1, a new kind of superbug never before seen in the United States. Medical professionals were alarmed for two reasons. One, this new superbug is resistant to colistin [Co—liss—tin], an antibiotic of last resort which is used when no other antibiotics can fight the infection.

Two, this MCR–1 gene can move from one bacteria to another. Eventually, MCR– 1 could merge with another superbug that is resistant to all antibiotics, except for colistin, and form an unstoppable superbug.

In response to the discovery of the MCR–1 gene, CDC Director Dr. Tom Frieden commented that "the medicine cabinet is empty for some patients . . . . It is the end of the road unless we act urgently." If the threat is not stopped, minor infections may become life threatening and treatment for disease such as cancer, diabetes or routine surgeries will be at risk.

Fortunately, the end of the road is not here yet. Congress and Federal agencies are working diligently to counter the effects of antibiotic resistance.

These efforts confront the two main contributors to the spread of antibiotic resistance: The over-prescription of antibiotics and the lack of new antibiotic development.

Since the discovery of penicillin in the early 20th century, almost every type of bacteria has become less responsive to antibiotics. As soon as an antibiotic goes in to wide use among the general public, bacteria evolve to become resistant.

A study published last month in the Journal of the American Medical Association (JAMA) found that nearly a third of antibiotics prescribed in doctors' offices, emergency rooms, and hospital-based clinics in the United States are not needed. This amounts to nearly 47 million unnecessary prescriptions given out each year.

And the numbers in this report most likely undercount the use of antibiotics, because the data did not include urgent care clinics, retail pharmacies, dentists' offices, and prescriptions given over the phone, and by nurse practitioners and physician assistants.

To combat antibiotic over-use, the CDC has partnered with FDA to advocate for antibiotic "stewardship" programs in health care facilities throughout the United States. The CDC has issued guidelines about how hospitals can minimize inappropriate or excessive use of antibiotics, which could help reduce antibiotic over-prescription.

Reducing the inappropriate use of antibiotics will help, but it can only slow the spread of antibiotic resistant bacteria. New antibiotics and alternative therapies must be developed.

Despite the need for antibiotic development, as of March 2016, there were only 37 new antibiotics in development. Just 13 were in phase 3 clinical trials. These drugs would potentially address many resistant bacteria—but they are not enough.

To combat this, in February of this year, the Biomedical Advanced Research and Development Authority (BARDA) has collaborated with NIH to establish a "Bio-pharmaceutical Accelerator" that will support research and development to incentivize antibacterial drug development. This Accelerator will (1) fund development of antibacterial products, (2) quickly move successful drug candidates through early development, (3) provide business and drug development guidance, and (4) decrease barriers to research and development of antibiotics.

The CDC, FDA, NIH, and BARDA have, and continue to make, significant and ongoing contributions to implement the National Action Plan for Combating Antibiotic-Resistant Bacteria released last year, which outlines steps to implement a national strategy to combat antibiotic resistance.

Additionally, Congress has increased funding for these initiatives by 57 percent over last fiscal year, for a total of more than $375 million.

Despite these promising developments, we are facing a public health challenge and we need to ensure that the federal government is taking the appropriate action to protect the American public.

I would like to thank the witnesses for appearing before the Subcommittee today and I look forward to hearing your testimony on this very important issue.

Mr. MURPHY. I now recognize the Ranking Member of the subcommittee, Ms. DeGette of Colorado, for 5 minutes.

And I now recognize the ranking member of the subcommittee, Ms. DeGette of Colorado, for 5 minutes.

Opening Statement of Hon. Diana DeGette, a Representative in Congress from the State of Colorado

Ms. DEGETTE. Thank you so much, Mr. Chairman. This is a really important hearing. We have been worried for quite some time here in Congress and at the agencies about the risk of resistant antibiotics. And this report last month about the superbug has been quite concerning to all of us.

The bacterium's resistance to colistin is particularly concerning because in this country physicians use colistin as the treatment of last resort when other antibiotics are no longer effective. Public health experts fear that this gene could jump to other bacteria that are already resistant to most other antibiotics.

Here is another observation about Dr. Frieden. He said, "It basically shows that the end of the road isn't very far away for antibiotics, that we may be in a situation where we have patients in our intensive care units or patients getting urinary tract infections for which we do not have antibiotics."

Obviously, we are all concerned about this, and I know all the witnesses and the members share this concern. We really don't want to revert to a time when physicians no longer have the tools to treat infections. Even common and once easily treated infections could once again prove life-threatening.

That this newest superbug has emerged on our home turf should not surprise us. Public health and infectious disease experts have been sounding the alarm for years. But I hope that this new discovery will lend urgency to efforts to monitor and fight antibiotic resistance.

The CDC has reported for decades that overuse of antibiotics and poor hygiene practices in hospitals and other inpatient health care settings contribute to the formation of drug-resistant bacteria. For example, studies showed that 30 to 50 percent of antibiotics prescribed in hospitals were unnecessary or inappropriate. Public health experts have similarly warned us about overprescribing of antibiotics in outpatient settings. A recent Pew study, for example, found that about 30 percent of all outpatient office visits in the U.S. resulted in the prescribing of an antibiotic, but 30 percent of those, which is almost 50, 5–0 million prescriptions are unnecessary.

Now, in the last decade these issues have received increased attention and funding, but as our witnesses will testify today, there is still far more to do. And I want to hear from our witnesses in particular in two different areas. The first one is antibiotic stewardship programs, which can both decrease the spread of infections and reduce the inappropriate use of antibiotics.

We need to improve public health education to ensure that patients and physicians understand how and when antibiotics should be prescribed. We need such stewardship programs both in health care facilities like hospitals and also in communities. I am hoping I can hear from our witnesses about antibiotic stewardship and whether we are seeing positive outcomes from current efforts.

I also am interested to see how antibiotic stewardship programs can result in more appropriate and effective use of antibiotics in animals. I think more needs to be done to shed a light on these issues, and I am interested to see what the witnesses will have to say. I do wish that we had a witness from the USDA because antibiotic overuse in animals is a really big problem.

The second area I want to hear about is the development of new antibiotics, diagnostics, and even vaccines to address the issue of antibiotic resistance. All of these agencies today play a critical role in the development of new drugs and other tools.

I can’t lose the opportunity to talk about our wonderful bill that Chairman Upton and I have cosponsored, along with all the other members of this subcommittee. Earlier this year, the 21st Century Cures Act passed out of the Energy and Commerce Committee unanimously and was then passed by the full House on an overwhelming basis. The bill includes the text of the Antibiotic Development to Advance Patient Treatment Act, or ADAPT Act. This is a really important bill that was originally cosponsored by Congressman Green and Congressman Shimkus, and it creates a new FDA approval pathway for limited population antibacterial drugs.

This legislation is designed to provide an approval process for drugs that affect a limited population of patients with serious or life-threatening infections or for drugs that fill an unmet need. We really need to pass the ADAPT Act. The easiest way to do this is of course for the Senate to pass the 21st Century Cures bill, and frankly, we keep hearing assurances that this will be happening any day. So maybe this urgent issue can be used to help enact this important law into law.

So, Mr. Chairman, I really want to thank you for having this hearing. It is an important one. And I want to thank our witnesses again for coming today, and I yield back.

Mr. MURPHY. The gentlelady yields back. I now recognize the chairman of the full committee, Mr. Upton, for 5 minutes.

OPENING STATEMENT OF HON. FRED UPTON, A REPRESENTATIVE IN CONGRESS FROM THE STATE OF MICHIGAN

Mr. UPTON. Well, thank you, Mr. Chairman.

So today, we gather to discuss the U.S. public health response to antibiotic resistance in light of the recent discovery of a new superbug. This superbug gene was first discovered by Chinese and British researchers in pigs, raw pork, and in a small number of folks in China in November of last year, but a recent case of course of the woman in Pennsylvania with the E. coli is the first discovery of this rare gene known as MCR here in the U.S.

A headline in the Post captured the urgency: “The superbug that doctors have been dreading just reached the U.S.” Concerns about this new threat are real indeed and they are being felt in Michigan and throughout the country.

The detection of this new antibiotic-resistant gene is very troubling because it signals the potential arrival of an unstoppable superbug. The gene is resistant to a last-report antibiotic and has the ability to move from one bacterium to another. While MCR–1 on its own is treatable by other antibiotics, disease experts tell us that the fear is not if but when this gene transfers and merges with another superbug that is resistant to all other antibiotics. This would create the nightmare scenario of a bacterial infection that cannot be stopped with any known antibiotic treatment.

The continuing evolution of bacteria, the overprescription of antibiotics, and the lack of new antibiotic development have all contributed to the problem. Our understanding of the MCR–1 gene is growing by the day, but there are still many questions that remain to be answered before we can be assured that we are doing everything that we can to protect the American people from this superbug and future challenges that arise from antibiotic resistance.

The questions include how did the bug get here to the U.S.? Where did it come from? How does it spread? Are we prepared for an outbreak of antibiotic-resistant bacterium? What is the Federal Government's plan to confront the public health challenge? In addition to our concerns about this particular superbug, we need to take a look at antibiotic resistance as a whole. And in response to the discovery of MCR gene in Pennsylvania, Dr. Frieden commented that it basically shows us that the end of the road isn't very far away for antibiotics. If that is true, minor bacterial infections could suddenly become fatal.

So we need to evaluate antibiotic development and assess where the science is, what barriers exist. How do we promote the discovery of new antibiotics? And as my good friend and colleague Ms. DeGette said, through the GAIN Act of 2012 and the ADAPT Act, which is part of 21st Century Cures, this committee has made a good number of strides to foster and encourage the development of new antibiotics that can fight the superbugs. In the meantime, we need to take appropriate measures such as antibiotic stewardship program to ensure that the antibiotics that already exist are being prescribed appropriately.

[The prepared statement of Mr. Upton follows:]

Opening Statement of Chairman Fred Upton Subcommittee on Oversight and Investigations Hearing on "Combatting Superbugs: U.S. Public Health Responses to Antibiotic Resistance," June 14, 2016

Today, we gather to discuss the U.S. public health response to antibiotic resistance in light of the recent discovery of a new superbug.

This superbug gene was first discovered by Chinese and British researchers in pigs, raw pork, and in a small number of people in China in November last year, but a recent case of a woman in Pennsylvania with E. coli is the first discovery of this rare gene, known as MCR-1, in the United States. A headline in *The Washington Post* captured the urgency – "The superbug that doctors have been dreading just reached the U.S." Concerns about this new threat are real, and they are being felt in Michigan and throughout the country.

The detection of this new antibiotic-resistant gene is troubling because it signals the potential arrival of an unstoppable superbug. This gene is resistant to a last-resort antibiotic and has the ability to move from one bacterium to another. While MCR-1 on its own is treatable by other antibiotics, disease experts tell us the fear is *not if – but when* – this gene transfers and merges with another superbug that is resistant to all other antibiotics. This would create the nightmare scenario of a bacterial infection that cannot be stopped with any known antibiotic treatment.

The continuing evolution of bacteria, the over-prescription of antibiotics, and the lack of new antibiotic development have all contributed to this problem. Our understanding of the MCR-1 gene is growing by the day, but there are still many questions that remain to be answered before we can be assured that we are doing everything that we can to protect the American people from this superbug and future challenges that arise from antibiotic-resistance.

The questions include: How did the bug get to the United States? Where did it come from? How does it spread? Are we prepared for an outbreak of an antibiotic-resistant bacterium? What is the federal government's plan to confront this public health challenge?

In addition to our concerns about this particular superbug, we need to take a look at antibiotic resistance as a whole. In response to the discovery of the MCR-1 gene in the Pennsylvania case, the CDC Director, Dr. Tom Frieden, commented that "[i]t basically shows us that the end of the road isn't very far away for antibiotics…" If that is true, minor bacterial infections could suddenly become fatal.

We need to evaluate antibiotic development and assess where the science is, what barriers exist, and how we can promote the discovery of new antibiotics. Through the GAIN Act in 2012, and the ADAPT Act which is part of 21st Century Cures, this committee has made numerous strides to foster and encourage the development of new antibiotics that can fight these superbugs.

In the meantime, we need to take appropriate measures, such as antibiotic stewardship programs, to ensure that antibiotics that already exist are being prescribed appropriately.

We thank the experts joining us this morning to discuss the federal response to this superbug and how antibiotic resistance is being addressed both as a nation and globally. The last thing we can afford, is looking back to today, and wishing we had done more.

Mr. UPTON. I appreciate the testimony and the dialogue with our experts, and I yield back to Mr. McKinley.

OPENING STATEMENT OF HON. DAVID B. MCKINLEY, A REPRESENTATIVE IN CONGRESS FROM THE STATE OF WEST VIRGINIA

Mr. MCKINLEY. Thank you, Mr. Chairman.

Antibiotics, it is my understanding—I am the engineer in the room, not the practicing physician, but my understanding is that the antibiotics are not effective against viruses, and this seems to be a good scientific fact from which to begin with. But yet in America you have heard the statistics. A third of the antibiotics are overprescribed by doctors. The CDC said that 47 million prescriptions were unnecessarily prescribed or inappropriately prescribed for cases that don't respond to the antibiotics.

And then the other point it comes down to is that we only represent 5 percent of the population in the world, so we have got 95 percent of the world out there—in many respects we know there are 20 nations in Europe and we know Mexico gives you antibiotics over the counter. And that tends to get to a point where we are probably going to overprescribe or they are going to be inappropriately used and our body will be able to build up resistance to that as a result of this inappropriate use for that.

And then we go to this, fast-forward to today with the woman in Pennsylvania. The interesting part is going to be how we are going to respond to that because how did she get it and what are we going to do to respond to that? So watching the time frame with this, we don't know if these antibiotics are used to help will even be effective in the future because of this overuse, overprescription.

[The prepared statement of Mr. McKinley follows:]

Opening Statement of Vice Chairman McKinley Subcommittee on Oversight and Investigations Hearing on "Combatting Superbugs: U.S. Public Health Responses to Antibiotic Resistance," June 14, 2016

Thank you, Mr. Chairman, and thank you for holding this hearing on this important public health challenge.

When we think about superbugs, or antibiotic resistant bacteria, we think about two different categories: what we know, and what we don't know.

We know that doctors and patients alike are partially to blame for the over-use of antibiotics. Doctors prescribe antibiotics when they are not necessary, and patients ask for antibiotics, thinking they will make them better, faster.

We know that antibiotics are becoming less effective in treating bacterial infections, because bacteria are evolving to become resistant. This will make it harder to treat not only common infections, such as strep throat, but also more serious conditions like cancer.

Patients fighting cancer undergo treatment that damages the immune system, which makes them more susceptible to infections. The same is true for organ transplant patients.

We know that the current crop of antibiotics in development and clinical trials are not sufficient. Researchers are starting to examine alternative therapies, other than antibiotics, to treat bacterial infections.

Now, on to what we don't know. We don't know how the dangerous MCR-1 gene came into the United States. We don't know how the woman in Pennsylvania contracted the MCR-1 gene, since she had not traveled outside the United States recently, and tests did not find the gene in her close family and friends.

We don't know how long it will take for the MCR-1 gene to transfer to bacteria that are resistant to all other antibiotics to create bacteria that cannot be treated with existing antibiotics. And when that happens, we don't know how far or how quickly it will spread, and how doctors will treat it.

Fortunately, Congress and Federal agencies are taking action, and they did not wait until this MCR-1 gene showed up. Over the last several years, the Federal government has ramped up its efforts to surveil and track these dangerous bacteria so that scientists and medical professionals can be as prepared as possible to confront these threats.

The CDC has led the coordinated effort between the DOD and the USDA to respond to the discovery of the MCR-1 gene. The USDA has also been investigating the source of the MCR-1 gene, and discovered the gene in a pig in the United States. The USDA is currently working to identify the source of that gene as well.

Through the National Antimicrobial Resistance Monitoring System, the FDA, CDC, and USDA all conduct research on bacteria found in food, animals, and humans. These surveillance methods track the bacteria and determine how resistance arises and transfers between bacteria.

Also, as part of a National Action Plan, CDC is ramping up its network of regional and local labs to track the spread of the MCR-1 gene and other new forms of antibiotic resistance.

The more we know about these superbugs – where they are, how they become resistant, and how they are transmitted to humans – the more we can prepare as a nation to combat these challenges.

Thank you to our witnesses for testifying today, and I look forward to a productive conversation.

Mr. MCKINLEY. So I am looking forward to your testimony today, and I yield back the balance of my time to Dr. Burgess.

Mr. BURGESS. I thank the gentleman for yielding. And I do want to thank our panelists for being here today. And I am most interested in knowing how you all are going to be working together on an interagency basis to see that we achieve the goals. I would like to hear from you perhaps some of your thoughts on what the legislative branch might do in addition to getting the Cures bill passed over in the other body.

And then finally, there was a significant amount of money made available in this area last December, and I would like to hear from each of you how that money has been allocated and utilized and as to whether or not you thought it had been of any benefit.

Thank you, Mr. Chairman. I will yield back.

Mr. MURPHY. Thank you. I now recognize the ranking member of the full committee, Mr. Pallone, for 5 minutes.

Opening Statement of Hon. Frank Pallone, Jr., a Representative in Congress from the State of New Jersey

Mr. PALLONE. Thank you, Mr. Chairman.

Antibiotic resistance is a significant domestic and global threat to both public health and national security, and I am glad we are taking the opportunity to examine this issue. I want to begin by emphasizing that antibiotics are incredibly valuable tools that have made once-fatal infections easily treatable, and antibiotics have transformed our health care system.

But as the use of antibiotics has spread, the threat posed by antibiotic-resistant bacteria or superbugs has grown. For that effective, coordinated, and decisive government action, we risk entering into a post-antibiotic world where common infections could once again become life-threatening.

The recent discovery of the MCR–1 gene on a colistin-resistant strain of the E. coli bacteria signals that this day is closer than it has ever been before. We see alarming statistics about the rates of overprescribing of antibiotics both for unnecessary and inappropriate use. At the same time, we have seen that drug manufacturers are unable to produce enough new drugs to meet this threat. There is no question that our arsenal of effective antibiotics is dangerously low today as a result of antibiotic resistance, and it is a dire situation.

That said, I am encouraged by the attention and funding we have placed on antibiotic resistance in recent years. Last year, the White House unveiled its National Action Plan for Combating Antibiotic-Resistant Bacteria, which sets forth ambitious goals to fight antibiotic resistance. All the agencies before us today play critical roles in that effort. In fiscal year 2016, we are devoting over $830 million to fighting antibiotic resistance, and President Obama has requested $1.1 billion towards this effort for fiscal year 2017. And these are important investments.

But in order for us to effectively address antibiotic resistance we need to make this a priority for the foreseeable future. This is not an issue we can address for a few years and then ignore. Antibiotic resistance is a reality of nature. Bacteria begin to mutate and develop resistance as soon as a new antibiotic begins to be used. We must develop long-term strategies and guarantee long-term funding to fight antibiotic resistance, and once we have begun implementing those strategies and ramping up funding, we must follow through on our commitments.

It is also important to emphasize that this is a global threat. It is not enough to address resistance here in the United States and hope that it will keep us protected. And from a global perspective, we must recognize that we are only as strong as our weakest link. In some places, antibiotics are used sparingly in the United States are available over the counter. In many countries, a lack of hygiene, sanitation, and basic infection control in medical settings results in increased antibiotic consumption.

If we do not want these superbugs here in the United States, we need to stop them from developing both here and abroad, and we must also improve global surveillance of antibiotic-resistant bacteria and antibiotic consumption. Stopping the spread of antibiotic-resistant bacteria is a global threat, and therefore, solutions must be global in nature.

So I just want to thank the witnesses, look forward to the promising work at each of your agencies and how we in Congress can be your partners in moving forward. And I yield my remaining time to Mr. Green of Texas.

Mr. GREEN. Thank you, Mr. Chairman. Thanks to our ranking member for yielding to me.

As said before, researchers have found a person in the United States carrying bacteria resistance to antibiotics of last resort, an alarming development that could mean the end of the road for antibiotics. For years, scientific leaders across the globe have been warning if we don't take swift, aggressive action, a post-antibiotic era in which modern medicine we take for granted is no longer safe because we will not be able to control infection from childbirth, in surgery, to dialysis and chemotherapy. Even simple cuts and wounds could be at increased risk of turning fatal without effective antibiotic treatments.

Alarmingly, at a time when resistance is rising, a pipeline for new antibiotics has dried up. Resistance can and must be slowed, but it cannot be stopped, and we need new classes of antibiotics to combat drug-resistant antibiotics.

The public value of antibiotics is difficult to overestimate, yet there is a widening acknowledged market fail when it comes to antibiotics. We need meaningful research incentives, strong public-private partnerships, and flexible clinical trials. That is why not only myself but our committee championed the ADAPT Act. It passed in the 21st Century Cures. It would enable the FDA to work with sponsors and reduce regulatory barriers to antibiotic drug development.

ADAPT offers one tool Congress can enact to address the lack of effective treatments against superbugs. I look forward to hearing more about that proposal and other ideas Congress should consider in the ongoing efforts to address this public health crisis.

And again, I will yield back my 14 seconds. Mr. MURPHY. The gentleman yields back.

I ask unanimous consent that the members' written opening statements be introduced into the record. And without objections, the documents will be entered into the record.

I would now like to introduce the witnesses of our first panel for today's hearing. The first witness on today's panel is Dr. Beth Bell. Dr. Bell is a Director of the National Center for Emerging and Zoonotic Infectious Diseases with the CDC.

We would also like to welcome again Dr. Janet Woodcock. Dr. Woodcock is currently Director of the FDA's Center for Drug Evaluation and Research.

Our third witness on today's panel is Dr. Richard Hatchett. Dr. Hatchett is currently serving as acting director for the Biomedical Advanced Research and Development Authority within HHS Office of the Assistant Secretary for Preparedness and Response.

Our final witness is Dr. Dennis Dixon. Dr. Dixon joins us from the NIAID's Division of Microbiology and Infectious Diseases.

I would like to thank all the witnesses for appearing before the subcommittee today. You are all aware that the committee is holding an investigative hearing, and when doing so has the practice of taking testimony under oath. Do any of you have an objection to taking testimony under oath?

The chair then advises that you are under the rules of the House and the rules of the committee you are entitled to be advised by counsel. Do any of you desire to be advised by counsel during the testimony today?

Seeing no request for that, in that case, will you all please rise, raise your right hand, and I will swear you in.

[Witnesses sworn.]

Mr. MURPHY. Thank you. All the witnesses have answered in the affirmative, and you are now under oath and subject to the penalties set forth in title 18, section 1001 of the United States Code.

I will have you each give a 5-minute summary of your written statement. Please pay attention to the timing and turn the microphone on and bring it as close to your mouth as you can. Thank you. You may begin, Dr. Bell. You are recognized for 5 minutes.

STATEMENTS OF DR. BETH BELL, DIRECTOR, NATIONAL CENTER FOR EMERGING AND ZOONOTIC INFECTIOUS DISEASE, CENTERS FOR DISEASE CONTROL; DR. JANET WOODCOCK, DIRECTOR, CENTER FOR DRUG EVALUATION AND RESEARCH, FOOD AND DRUG ADMINISTRATION; DR. RICHARD J. HATCHETT, ACTING DIRECTOR, BIOMEDICAL ADVANCED RESEARCH AND DEVELOPMENT AUTHORITY; AND DENNIS M. DIXON, DIVISION OF MICROBIOLOGY AND INFECTIOUS DISEASES, NATIONAL INSTITUTE OF ALLERGY AND INFECTIOUS DISEASES, NATIONAL INSTITUTES OF HEALTH

Statement of Dr. Beth Bell

Dr. BELL. Thank you. Good morning, Chairman Murphy, Ranking Member DeGette, and members of the subcommittee. I am Dr. Beth Bell, director of the National Center for Emerging and Zoonotic Infectious Diseases at the Centers for Disease Control and Prevention. And thank you for the opportunity to testify before you today.

CDC works 24/7 to save lives and protect people against health threats, including the threat of potentially untreatable infections. With support from

Congress, CDC is working to improve the Nation's capacity to detect, respond, and prevent antibiotic-resistant threats in health care settings and across communities to protect Americans and save lives.

Using the resources provided by Congress in fiscal year 2016, CDC is working on four fronts to support transformative improvements in our national capability to identify and respond to antibiotic resistance. First, CDC will invest the largest portion of this funding in our 50 states, six largest cities, and Puerto Rico. CDC will support state activities to improve the detection and prevention of AR infections transmitted across health care and community settings. These are just-in-time investments as the pace and challenges of antibiotic resistance are accelerating and every state needs capacity to take appropriate action.

Second, CDC's antibiotic resistance lab network will provide infrastructure and lab capacity through seven regional labs across the country. These labs will be able to detect resistant organisms recovered from human samples and new forms of antibiotic resistance. CDC will also provide support for labs in all States to test for CRE, the nightmare bacteria.

Third, CDC's Advanced Molecular Detection initiative is another important tool in our efforts to identify and solve more outbreaks faster. Because of innovations developed through this AMD initiative, CDC will be able to scale up whole genome sequencing of multiple foodborne pathogens to better understand foodborne antibiotic resistance patterns. Through the National Antimicrobial Resistance Monitoring System, NARMS, sequences from animals, from retail meat, and from humans will be compared to identify new ways of preventing human infections.

Finally, our ability to reduce antibiotic resistance will depend in part on improving antibiotic use, and we are working towards that through better measurement, through expansion of stewardship programs, and through educational campaigns.

The administration's budget request for fiscal year 2017 includes an increase of $40 million for year 2 of CDC's Antibiotic Resistance Solutions initiative. And so in fiscal year 2017, in addition to sustaining AR capacity started in '16, CDC will expand state antibiotic resistance prevention programs to better respond to outbreaks, to improve prescribing, and prevent antibiotic-resistant infections across all health care settings.

These collective investments have a direct impact on the response to AR threats, including the emergence of the MCR–1 gene in the United States. In May, Department of Defense scientists announced the first discovery of the MCR–1 gene in bacteria isolated from a person in the United States. Although there is not an immediate threat to the public or the current health of this patient, this is an important development for the United States. The antibiotic colistin is used as a rescue drug to treat patients with multidrugresistant infections like CRE.

The MCR–1 gene makes bacteria resistant to colistin. The gene exists on a plasmid, which is a small piece of DNA that is capable of moving from one bacterium to another, spreading antibiotic resistance among bacterial species. The presence of the MCR–1 gene and its ability to share its colistin resistance with other bacteria such as CRE raises the possibility of a bacteria that is resistant to every antibiotic.

USDA also discovered MCR–1 in E. coli isolates collected from two different pig intestines. By comparing the DNA sequences of all three isolates, Federal scientists have determined that the isolates from the pigs are different from that from the human.

Yesterday, CDC issued an alert to proactively notify states, hospitals, and clinical laboratories about the availability of new detection tools for MCR–1 and to reiterate recommendations for infection prevention, environmental cleaning, and reporting to public health.

The identification of MCR–1 vividly illustrates our domestic and global challenges of antibiotic resistance. MCR–1 was first identified in China in November of last year, and in less than 6 months has been identified in a human and two animals in the United States. Antibiotic use anywhere can potentially affect any one of us.

The emergence and reemergence of health threats, including those caused by antibiotic-resistant bacteria, is something we can expect to continue to see in the future. If we lose antibiotics, we could lose the ability to effectively treat sepsis and to provide the care to cancer patients, to organ transplant recipients, or burn and trauma victims.

So thank you again for the opportunity to appear before you today, and I look forward to the opportunity to answer your questions.

[The prepared statement of Beth Bell follows:]

Witness: Beth Bell, MD, MPH; Testimony: House Energy and Commerce Committee, Subcommittee on Oversight and Investigations

Good morning Chairman Murphy, Ranking Member DeGette, and members of the Subcommittee. Thank you for the opportunity to testify before you today on the Centers for Disease Control and Prevention's (CDC's) continued efforts to combat antibiotic resistance, which threatens the United States and modern medicine itself.

CDC is the nation's premier health protection agency, working 24-7 to save lives and protect people against the threat of untreatable infections. We know that, although much effort is being expended to develop new antibiotics, we also must work right now to slow the spread of these resistant bacteria and improve how we use the antibiotics we have.

Threat of Antibiotic Resistance

Antibiotic resistance is perhaps the single most important infectious disease threat of our time. Every year, more than two million people in the United States get infections that are resistant to antibiotics, and at least 23,000 people die as a result. In addition, Clostridium difficile (C. difficile), a serious diarrheal infection associated with antibiotic use, causes at least 15,000 deaths every year in the United States.

Modern medicine is at stake. If we lose antibiotics, we lose the ability to effectively treat sepsis and to provide care to cancer patients, organ transplant recipients, and burn or trauma victims. Losing antibiotics would devastate our medical system. In 2014, the President issued an Executive Order directing the development of the National Action Plan for Combating Antibiotic Resistant Bacteria to identify steps Departments across the Federal Government could take to combat antibiotic resistance.

CDC's Antibiotic Resistance Solutions Initiative

In fiscal year (FY) 2016, Congress recognized the large and growing threat of antibiotic resistance (AR) and appropriated $160 million to CDC to implement the National Action Plan for Combating Antibiotic-Resistant Bacteria (CARB) activities: To detect and respond to resistant pathogens; prevent the spread of resistant infections; and collaborate with partners to encourage innovation for new prevention strategies. These resources are transforming how our nation tackles and slows antibiotic resistance comprehensively, efficiently, and systematically.

With support from Congress, CDC is working with Federal, state, and local public health, academic, healthcare, and veterinary partners to: improve capacity to detect and respond to AR threats in healthcare and communities, protect patients, and save lives. CDC will invest the largest extramural portion of this funding in the 50 state health departments, the six largest local health departments, and Puerto Rico. This Antibiotic Resistance Solutions Initiative will support comprehensive and coordinated public health action to minimize the spread of antibiotic resistance across states, counties, and cities.

Detect and Respond

This public health investment opens a new chapter against antibiotic resistance, improving state health department healthcare-associated infection (HAI) and AR detection capacity so they can better detect resistance and respond to outbreaks faster. Beginning this fall, CDC's Antibiotic Resistance Lab Network will provide infrastructure and lab capacity through as many as eight regional labs across the country. These labs will be able to detect resistant organisms recovered from human samples and new forms of antibiotic resistance—including mutations that allow bacteria to withstand last-resort drugs like colistin—and report these findings to CDC as well as back to facilities and states. These efforts will generate better data for stronger infection control to contain current threats and prevent future resistance threats.

CDC is particularly concerned about carbapenem-resistant Enterobacteriaceae (CRE)—often referred to as the nightmare bacteria—and emerging multidrug-resistant organisms. In addition to building on existing HAI/AR state programs, CDC will provide support for labs in all states to test for CRE and support regional labs for additional testing in outbreak response (e.g., screening colonized patients). The Emerging Infections Program (EIP), a national resource for surveillance, prevention, and control of emerging infectious diseases, will also build on population-based detection of all invasive Staphylococcus aureus (including MRSA), carbapenem-resistant Pseudomonas, and ESBL-producing Enterobacteriaceae.

CDC's advanced molecular detection (AMD) initiative is another important tool in our efforts to identify and address antibiotic resistance. The expansion of whole genome sequencing into public health surveillance activities represents a paradigm shift that has already saved lives by identifying and solving more outbreaks, faster. For example, CDC is using AMD to better characterize Listeria bacteria in foodborne disease outbreaks. Earlier this year, whole genome sequencing was used in a multi-state Listeria outbreak associated with packaged salads; this technology confirmed that cases identified in Canada were closely related. CDC will scale up whole genome sequencing of multiple foodborne pathogens—including Salmonella, Shigella, and Campylobacter—to better understand foodborne AR patterns. CDC will also use the National Antimicrobial Resistance Monitoring System (NARMS) and work with the Food and

Drug Administration (FDA) and the U.S. Department of Agriculture (USDA) to sequence foodborne pathogens to identify new ways of preventing human infection associated with resistant bacteria in food animals. Finally, CDC will provide states with additional epidemiologic support for foodborne investigations to ensure rapid response to new illness reports and outbreaks of resistant infections identified through this new testing.

Prevent

CDC's Antibiotic Resistance Solutions Initiative will help slow resistance and protect the antibiotics we have today with active coordination of care across communities, data collection, and stewardship program implementation. State HAI/AR Prevention Programs will work with healthcare facilities in up to 25 states to scale up this coordinated approach to prevent the transmission of multidrug-resistant organisms and C. difficile across communities.

Our ability to reduce antibiotic resistance will depend in part on improving antibiotic use, and we are working toward that through better measurement of antibiotic use, expansion of stewardship programs, and support for educational campaigns. CDC is working with large healthcare systems to demonstrate best practices in stewardship programs using CDC's Core Elements of Antibiotic Stewardship and measure antibiotic use outcomes. Using electronic data through the National Healthcare Safety Network's (NHSN's) antibiotic use module, we can assess appropriate use and target interventions in and across healthcare settings. These data will inform how facilities can improve antibiotic use and reduce resistance. CDC is collaborating with partners including the Centers for Medicare & Medicaid Services (CMS), Quality Improvement Networks (QINs), and CMS's Center for Medicare and Medicaid Innovation (CMMI) to implement effective stewardship programs.

Much of this work in turn responds to the threat of sepsis, often a sign of severe infection that can include resistant infections. CDC is committed to preventing infections that lead to sepsis and promoting early recognition of sepsis across medical care.

Finally, CDC will also work with FDA and USDA to ensure that practicing veterinarians have the tools, information, and training to prevent drug resistance by promoting responsible use of antibiotics in animals, including those animals consumed by humans.

Innovate

CDC's Antibiotic Resistance Solutions Initiative will accelerate prevention innovation, giving us new tools to slow the threat of resistance. Through the Prevention Epi-Centers and other innovation partners, CDC is able to address questions and gaps in knowledge related to resistant infections. CDC is also performing intramural studies and supporting extramural studies to better understand the microbiome and how its disruption influences an individual's risk for developing an antibiotic resistant infection. This work could help in tailoring prevention and antibiotic stewardship programs and uncovering new therapies that restore the healthy microbiome, reducing the risk of AR infections.

Collaborate

CDC is working with state, Federal, and international partners to align AR activities and maximize the effectiveness of funding. For example, CDC's efforts in whole genome sequencing for foodborne AR infections will involve collaborations with the FDA, USDA, and the National Institutes of Health (NIH) to track how antibiotic resistant genes in pathogenic and commensal bacteria that may be impacting humans, food, and animals. CDC, NIH, and the Department of Defense (DoD) are working together on when and how to best use whole genome sequencing to solve CRE outbreaks. CDC is collaborating with FDA on supporting and expanding use of the AR Isolate Bank, including coordinating efforts to generate genomic sequence data for all isolates in the AR Isolate Bank, which supports industry and academic partners working to develop new treatments and diagnostics.

Additional Investments

The Administration's budget request for FY 2017 includes an increase of $40 million for year two of CDC's Antibiotic Resistance Solutions Initiative. This increase will build on AR capacities started in FY 2016, allowing CDC to expand the nation's ability to detect, respond to, and prevent AR infections across healthcare settings and in the community.

In FY 2017, in addition to sustaining AR capacities started in FY 2016, CDC will expand the State HAI/AR Prevention Programs in up to 50 states, six large cities, and Puerto Rico to better respond to outbreaks, improve prescribing, and prevent AR infections across all healthcare settings (e.g., inpatient, outpatient, and long-term care settings). CDC will also further expand state public health laboratory capacity in up to 50 states, six large cities, and Puerto Rico to rapidly screen enteric bacteria for resistance and ensure the nation's ability to rapidly detect and investigate AR threats across the country and in more enteric pathogens, specifically Campylobacter and Shiga toxin-producing E. coli.

These collective investments have a direct impact on the public health response to the AR threat, including our continued response to new threats, such as the emergence of the mcr-1 gene, reported in November of 2015 in China.

Public Health Response to Discovery of mcr-1

The antibiotic colistin is used as a last-resort drug to treat patients with multidrug-resistant infections, including CRE. The mcr-1 gene makes bacteria resistant to colistin. The gene exists on a plasmid, a small piece of DNA that is capable of moving from one bacterium to another, spreading antibiotic resistance among bacterial species. CDC and Federal partners have been hunting for this gene in the U.S. since its emergence in China was reported in 2015.

Following the identification of mcr-1 in China, CDC, FDA and USDA began searching for mcr-1 in bacterial samples taken from human, retail meat, and food animal sources. This work is facilitated by the National Antimicrobial Resistance Monitoring System (NARMS), which has detected emerging resistance to clinically important antibiotics for the past 20 years. NARMS is a partnership among FDA, USDA, CDC, and state and local public health departments that tracks changes in the antimicrobial susceptibility of intestinal bacteria. Through NARMS, the USDA discovered mcr-1 in an E. coli isolate collected from a pig intestine. The DNA sequence of that isolate revealed that the strain contained the mcr-1 gene on a plasmid. USDA scientists also determined that the mcr-1 carrying colistin-resistant E. coli from intestinal content of a pig was resistant to other antibiotics, including ampicillin, streptomycin, sulfisoxazole, and tetracycline. The resistance to these other antibiotics was not on the plasmid carrying the mcr-1 gene. Preliminary analysis by USDA's Agricultural Research service indicates that the bacterial strain is E. coli 0160:H40. An isolate from a second pig is undergoing analysis. As of April 2016, more than 55,000 genome sequences of several types of bacteria from humans, retail meat, and food animals have been screened through NARMS. All were negative for the mcr-1 gene. NARMS continues to look for this gene.

In addition, CDC investigated bacterial genomes and used a highly sensitive testing method known as polymerase chain reaction, or PCR-screening, to look for the mcr-1 gene among healthcare-associated pathogens. In this effort, CDC screened 735 genomes, including 690 Enterobacteriaceae (e.g., carbapenem-susceptible Enterobacteriaceae (CSE) and carbapenem-resistant Enterobacteriaceae (CRE)) and 45 non-Enterobacteriaceae (e.g., Pseudomonas and Acinetobacter) from the surveillance, outbreak, and special study and reference collections. All these human isolates were negative for mcr-1 using this PCR screening method.

In May 2016, DoD scientists announced the first discovery of the mcr-1 gene in bacteria isolated from a person in the United States. CDC is currently part of the coordinated public health response to DoD's identification of the gene in E. coli bacteria in a urine sample from a Pennsylvania woman with no recent travel outside of the country. The isolates identified from the 1st pig is different from the isolate identified in the human case; the isolate from the 2nd pig is currently undergoing analysis.

The DoD finding of the mcr-1 gene in bacteria from a person started an ongoing public health investigation led by CDC and the Pennsylvania Department of Health. Although only one human isolate has been recovered in the United States to date, CDC takes this finding very seriously and is working with DoD, the Pennsylvania Department of Health, local health departments, and others to identify people who have had contact with the Pennsylvania patient. The investigation is currently focused on identifying and screening close contacts, including household and healthcare contacts of the Pennsylvania patient to determine whether any of them might carry bacteria containing the mcr-1 gene.

Despite some media reports, the Pennsylvania Department of Health investigation has determined that the woman did not have CRE, and the bacteria identified is not resistant to all antibiotics (referred to as a pan-resistant infection). The presence of the mcr-1 gene, however, and its ability to share its colistin resistance with other bacteria, such as CRE, raises the possibility that pan-resistant bacteria could develop.

The transformative improvements being implemented across the country through CDC's Antibiotic Resistance Solutions Initiative will greatly expand and strengthen national efforts to identify and respond to this gene if it is circulating.

Conclusion

The emergence and reemergence of health threats, including those caused by antibiotic resistant bacteria, is something we can expect to continue to see in the future. Consistent with the National Action Plan for Combating Antibiotic-Resistant Bacteria, CDC and other government agencies continue to track the emergence of antibiotic resistance. CDC is working to strengthen the nation's ability to respond rapidly and effectively, protecting Americans and the precious antibiotics we need to fight deadly bacteria.

Thank you again for the opportunity to appear before you today. I appreciate your attention to the antibiotic resistance threat, and I look forward to answering your questions.

Mr. MURPHY. Thank you. Dr. Woodcock, you are recognized for 5 minutes.

STATEMENT OF DR. JANET WOODCOCK

Dr. WOODCOCK. Thank you for holding a hearing on this critical topic.

FDA is involved in many fronts in the fight against resistant organisms. We're participating actively in the CARB initiative and collaborating internationally and also of course with our Federal partners represented here. Our Center for Veterinary Medicine oversees animal drugs. They have received commitments from the manufacturers are medically important antibiotics to submit supplements by the end of the year. As a result of this, these drugs will have to be prescribed by a veterinarian and can't be used for growth promotion purposes.

Our Center for Devices and Radiologic Health regulates in vitro diagnostics or test kits, including antimicrobial susceptibility tests and diagnostics. They're seeing a growing pipeline of rapid diagnostic tests under development. This is very good news. These tests are intended to identify if a person has a bacterial infection or— versus a virus or has bacteria circulating in the blood or even rapidly diagnose certain types of resistant organisms. So we hope these development programs are successful.

They also plan to issue guidance on what they call coordinated development of susceptibility tests. So as we have new antimicrobials approved by Center for Drugs, we could have susceptibility tests also available that could be interpreted by clinicians.

Our Center for Biologics regulates vaccines. We know that prevention is a best approach to disease, and of course there are some vaccines for microbial diseases, and we would hope that more are going to be developed.

The Center for Drug Evaluation and Research that I head regulates antimicrobials. We've seen the percent of infections treated by common antibiotics shrink over time, as everybody's been saying, as resistance grows. CDER's been actively implementing the GAIN Act that Congress passed several years ago. We've granted 107 qualified infectious disease product designations to 63 different molecules. The designated products can receive fast-track designation, and they get a priority review. On approval, they receive additional 5 years of market exclusivity, so this is an incentive of course. Five designated antibacterial drugs and one antifungal drug have been approved since GAIN was enacted.

We continue to work on streamlining drug development. We've issued guidelines. We're trying to lower the barriers, but many of the barriers, as I'm sure we'll discuss, are either scientific or economic barriers to new antimicrobials.

Now, there are further efforts going on in lowering the barriers. I really commend BARDA who is exploring the use of a master common protocol. We're working with them, novel ways of studying these that lower the barriers and make some types of studies actually feasible.

The 21st Century Cures bill that was passed in the House contains provisions for the limited population use provision, and the Senate PATH Act does mirror that. We're also very interested in the issue of antimicrobial break points and the use of a more rapid and agile method for changing the break point designations.

Finally, though, drug development in this area remains fragile and weak. The incentives that have been put in place apparently are not enough to overcome the scientific challenges that the industry faces in finding new targets and developing these and then actually making money on them. Much more needs to be done, I think, to address this threat across all the groups represented here, but the drug development area in particular needs further examination.

Thank you.

[The prepared statement of Janet Woodcock follows:]

Statement of Janet Woodcock, M.D., Director, Center for Drug Evaluation and Research Food and Drug Administration, Department of Health and Human Services, before the Subcommittee on Oversight and Investigations Committee on Energy and Commerce, U.S. House of Representatives, "Examining Ways to Combat Antibiotic Resistance and Foster New Drug Development," June 14, 2016 Release Only upon Delivery

Introduction

Mr. Chairman, Ranking Member DeGette, and Members of the Subcommittee, I am Dr. Janet Woodcock, Director of the Center for Drug Evaluation and Research (CDER) at the Food and Drug Administration (FDA or the Agency), which is part of the Department of Health and Human Services (HHS). Thank you for the opportunity to be here today to discuss the current state of antibiotic resistance and the need for new solutions to the current crisis.

To address the growing public health concern, on September 18, 2014, the Administration released the National Strategy for Combating Antibiotic-Resistant Bacteria (CARB), and President Barack Obama signed Executive Order 13675, which called for creation of a National Action Plan and established a Federal Task Force to draft and implement the action plan. FDA is actively engaging in these efforts in coordination with other Federal agencies.

The decline in antibacterial drug research and development (R&D) in the private sector, at a time when serious antibiotic resistant infections are on the rise, is a tremendous public health problem, resulting in a very serious unmet medical need. According to the Centers for Disease Control and Prevention (CDC), each year in the United States, at least two million people become infected with bacteria that are resistant to antibiotics and at least 23,000 people die each year as a result of these infections.

Antibacterial drugs are critically important across medicine, including in the care of premature infants and for use in surgery, chemotherapy, and organ transplantation. However, bacteria are adept at becoming resistant to antibacterial drugs, as highlighted by the recent identification of a patient in Pennsylvania infected with *E. coli* bacteria possessing the *mcr-1* gene. It is essential to use antibiotics judiciously to slow the development of resistance. Moreover, new antibacterial drugs are needed to provide treatment options in cases where resistance has eroded the effectiveness of existing drugs.

The Challenges Impacting Antibacterial Drug Development

There are significant scientific and economic challenges impeding the development of new antibiotics. From a scientific standpoint, many patients with bacterial infections are very sick and need to begin antibiotic therapy immediately. However, enrolling a very sick patient in a clinical trial in order to evaluate new antibiotics at the same time they are very sick can be difficult because critically ill patients may be unable to provide informed consent.

From an economic standpoint, antibiotics are generally viewed as less profitable by companies and venture capitalists because of their relatively low price and because they are generally taken only for a short period of time and often only for one course of treatment by any given patient. Compare this to the long, dependable income stream from a diabetes medicine or a blood pressure medicine that patients often take for the rest of their lives, or the relatively high price associated with cancer and some antiviral drugs. These economic realities can make it challenging for a company to justify large expenditures for the development of drugs in this area, as a report from the Eastern Research Group (ERG), funded jointly by HHS Office of the Assistant Secretary for Planning and Evaluation (ASPE) and FDA, affirms.[1]

Furthermore, the inappropriate use of antibacterial drugs can accelerate the development of antibiotic resistance. It is essential that we use antibiotic drugs prudently in order to preserve the effectiveness of these drugs. The ability of drug resistance to be transferred from one bacteria to another and spread among a population of patients is a phenomenon unique to infectious diseases. Judicious use of antibacterial drugs is essential, in both the human and animal sectors.

Use of Common Clinical Trial Protocols to Encourage Antibacterial Drug Development

A promising strategy for encouraging antibacterial drug development is the establishment of a clinical trial network that can operationalize common clinical trial protocols at a level of quality that matches the pharmaceutical industry and that can respond to future needs in an agile fashion. The quality of the network should be at a level where pharmaceutical companies welcome the opportunity to perform their core development program utilizing the common protocols that are housed within the clinical trial network. Such a clinical trial network would allow pharmaceutical companies to utilize shared expertise and infrastructure to study antibacterial drugs. The CARB calls out this specific approach to reducing obstacles faced by drug companies developing new antibacterial

---

[1] *http://aspe.hhs.gov/sp/reports/2014/antibacterials/rpt_antibacterials.cfm*.

drugs and states that the U.S. Government will examine the feasibility of generating and applying master clinical protocols to multiple test groups of patients while sharing a common control group. Colleagues at the Office of the Assistant Secretary for Preparedness and Response's Biomedical Advanced Research and Development Authority (BARDA) recently published a request for information on an antibacterial drug clinical trial network to gather information to evaluate the costs and practical considerations involved in establishing such a network.

What FDA Is Doing to Address the Current Challenges

Provisions in a law passed in 2012, commonly known as the Generating Antibiotics Incentives Now Act, or the GAIN Act, are helping to stimulate the development of new antibiotics. Under GAIN, certain antibacterial or antifungal drugs intended to treat serious or life-threatening infections can be designated as "Qualified Infectious Disease Products" (QIDPs). As part of its QIDP designation, a drug receives priority review[2] and is eligible for fast-track designation.[3] At the time of approval, a product with QIDP designation qualifies for an additional five years of marketing exclusivity, to be added to certain existing exclusivity periods under the Federal Food, Drug, and Cosmetic Act. To date, FDA has granted 107 QIDP designations for 63 different unique molecules.

Since GAIN was passed, FDA has approved five new antibacterial drugs and one new antifungal drug with the QIDP designation. Three of the five antibacterial drugs, Dalvance (dalbavancin), Orbactiv (oritavancin), and Sivextro (tedizolid phosphate), are intended to treat

---

[2] Priority-review designation directs overall attention and resources to the evaluation of applications for drugs that, if approved, would be significant improvements in the safety or effectiveness of the treatment, diagnosis, or prevention of serious conditions or for drugs that have a QIDP designation. Priority-review designation does not affect the length of the clinical trial period. FDA informs the applicant of a priority-review designation within 60 days of the receipt of the original BLA, NDA, or efficacy supplement.

[3] Fast-track designation is a process designed to facilitate the development and expedite the review of drugs to treat serious conditions and fill an unmet medical need. Once a drug receives fast-track designation, early and frequent communication between the FDA and a drug company is encouraged throughout the entire drug development and review process.

acute bacterial skin and skin-structure infections (ABSSSI) caused by methicillin-resistant *Staphylococcus aureus* and certain other types of bacteria. Zerbaxa (ceftolozane/tazobactam) is indicated for the treatment of complicated urinary tract infections (cUTI) and complicated intra-abdominal infection (cIAI). Avycaz (ceftazidime/avibactam) is indicated for cUTI and cIAI for patients who have limited or no alternative treatment options.

FDA is working hard to streamline requirements for clinical trials for studying new antibacterial drugs, and the provisions of the GAIN Act are being actively implemented, but much more is needed. There are still significant economic and scientific challenges in the development of new antibacterial drugs that need to be addressed. Additional approaches to reducing the costs of studying antibacterial drugs, such as common clinical trial protocols, could provide other important means to stimulate antibacterial drug development. We also need cutting-edge science to move forward the development of new and innovative antibacterial drugs, as well as alternative therapeutics to combat bacterial infections.

To help drive this effort, FDA is working with academia, regulated industry, professional societies, patient advocacy groups, and government agencies to advance the science of clinical trials for antibacterial drugs. For example, FDA has contributed to the efforts of the Biomarkers Consortium of the Foundation for the National Institutes of Health (FNIH) to develop new endpoints for studying antibacterial drugs. FDA also works closely with the Clinical Trials Transformation Initiative (CTTI), a key group of dedicated scientists focused on streamlining and advancing clinical trials for more efficient drug development. FDA and CTTI have partnered to initiate activities related to efficient design and conduct of clinical trials for testing new antibiotics. FDA has also teamed up with colleagues at the Brookings Institution's Engelberg Center for Health Care Reform and more recently the Duke-Margolis Center for Health Policy to galvanize the scientific community's efforts in new antibiotic drug development. The first Brookings Council for Antibacterial Drug Development (BCADD) meeting was held in August 2012, and the Brookings Institution has continued to convene meetings focused on a range of antibacterial drug development issues.

As part of these collaborative efforts, FDA will conduct a two-day public workshop in July 2016 focused on facilitating antibacterial drug development for patients with unmet needs and developing antibacterial drugs that target a single species. Discussions will focus on potential development pathways, aspects of clinical trials including which patient populations to study, trial designs and endpoints, and the role of clinical trial networks in antibacterial drug development.

Encouraging the Development of New Antibacterial Drugs

Given the public health threat posed by antimicrobial resistance, it is necessary to consider new mechanisms for encouraging the development of new antibacterial drugs to address unmet medical needs in the treatment of serious and life-threatening bacterial infections. We look forward to ongoing engagement with consumers, clinical experts, researchers, industry, and others to achieve this goal.

As the Committee knows, one option that has been proposed is the establishment of a new Limited Population Antibacterial Drug (LPAD) program. It is our understanding that, as a general matter, drugs approved using an LPAD pathway would be based on more streamlined development programs that establish that the drug is safe and effective in a limited population of patients with serious or life-threatening infections and unmet medical needs.

Importantly, because LPAD drugs would be approved based on streamlined development programs, there would be more uncertainty about potential risks posed by the product. This may result in a positive benefit-risk profile in a limited population of patients with serious or lifethreatening infections and unmet medical needs. However, the benefit-risk assessment would be different for a broader, more heterogeneous patient population with less serious manifestations of the infection and which has other treatment options. This distinction should be clearly conveyed to the provider community. A clear branding mechanism would convey accurately to physicians using the product the limitations of the data supporting approval, including the uncertainty and the unique benefit-risk profile associated with the drug. Such labeling is particularly important in the context of antibiotic drugs, where historical overuse has led to increased antimicrobial resistance.

Accelerating Development of Rapid Diagnostic Tests for Antibiotic Resistance

Rapid diagnostic tests are an essential component of combating antibiotic resistance. Specific goals for accelerating the development of these rapid diagnostics are identified in the CARB. Rapid diagnostics can aid in faster diagnosis of patients with antibiotic resistant infections and result in appropriate therapy being initiated earlier. Rapid diagnostics also serve an important role in identifying patients colonized with resistant isolates at entry to hospitals, such that patients can be isolated and transmission to other patients minimized. Perhaps even more importantly, rapid diagnostics may aid in differentiating viral infections from bacterial infections and substantially reduce the use of unnecessary outpatient antibiotic therapy.

Through release of guidance documents and several public workshops, FDA and industry have made great progress in this area, however, similar to drug development, there are significant challenges to the development and use of diagnostics relative to empiric antibiotic use. Strategies to incentivize development and use of new antibacterial drugs have similar parallels for diagnostics.

Expedited Updating of Susceptibility Test Interpretive Criteria (Breakpoints) to Maximize the Effective Use of Antimicrobial Products

Enabling physicians to select appropriate antibacterial drugs is critical to individual health, as well as the public health, as we continue to combat antimicrobial resistance. Generally, physicians rely on antimicrobial susceptibility test (AST) devices, which provide information about whether a bacterium is either susceptible or resistant to an antibacterial drug. The criteria used to determine susceptibility are commonly referred to as "breakpoints." This information helps physicians choose appropriate antibacterial drugs for treatment. The results of such testing is also important to identify patients with certain resistant bacteria that warrant additional infection control measures to prevent spread of their resistant bacteria to others.

Outdated breakpoints can cause numerous problems, such as interfering with the implementation of appropriate infection control procedures, which can increase the chances that problematic resistant bacteria can be spread to others. Hospitals need up-to-date breakpoint information in order to determine whether an infection is caused by a resistant pathogen, and to put appropriate infection control procedures in place for those antibiotic-resistant bacteria.

AST device manufacturers need to be able to incorporate up-to-date breakpoint information into their devices quickly. However, currently, it can take several years to do so. Under the current regulatory framework, each antibacterial drug manufacturer updates its drug labeling with new breakpoint information and only then does each device manufacturer update its device algorithms and labeling. Reviewing breakpoint labeling supplements for each individual drug product (even when it shares the same active ingredient(s), and thus, generally has the same breakpoints) is no small task. Moreover, the process begins with the submission of labeling supplements from the drug manufacturers.

This protracted process of manufacturers updating the product labeling for each antimicrobial drug product adversely affects the public health by delaying AST device manufacturers from being able to promptly update the breakpoint information in their devices. It utilizes both industry and Agency resources that could otherwise be used for antibacterial and antifungal drug development or reviews that could confer greater benefits for patients. We need a better, more modern and streamlined process to help AST device manufacturers incorporate up-to-date and comprehensive breakpoint information in their devices more quickly, in order to get this information to health care providers sooner for the care of patients.

Solution for Updating Breakpoint Information Faster

In order to address the problems with the current scheme for updating breakpoints, we need to take breakpoints out of the drug product label and utilize more rapid, electronic means of communicating this information. Posting breakpoint information on FDA's website could enable us to update breakpoint information more efficiently. As mentioned, many antibacterial drugs have the same active ingredient(s), and thus the same breakpoints. Accordingly, as a general matter, breakpoints are neither proprietary, nor specific to a particular drug product. Therefore, if FDA posted appropriate breakpoints for penicillin or amoxicillin products on the Internet, then FDA could take one single action to update the breakpoints for multiple drug products with the same active ingredient simultaneously.

To help FDA ensure that it can update breakpoint information accurately and expeditiously, the Agency could leverage the work being done by standards-development organizations that develop breakpoints, and recognize them, when FDA agrees that they are appropriate. FDA would retain full authority to accept a standard in whole or in part, or to establish alternative breakpoints. In addition, companies could submit data to support alternative breakpoints, if they disagree with the recognized standard.

Conclusion

It is virtually undisputed that we are facing a public health crisis because of the rise of serious resistant infections and the simultaneous decline in R&D in this area. FDA is using the tools we have to begin to strengthen the antibiotic drug pipeline. However, more work is needed to improve the current climate, and FDA is looking forward to continuing to work with stakeholders to address this public health crisis.

I am happy to answer any questions you may have.

Mr. MURPHY. Thank you, Dr. Woodcock.

I now recognize Dr. Hatchett for 5 minutes.

STATEMENT OF DR. RICHARD J. HATCHETT

Dr. HATCHETT. Good morning, Chairman Murphy, Ranking Member DeGette, and members of the committee. Thank you for inviting me to testify about our efforts to combat antibiotic-resistant bacteria. I am Dr. Richard Hatchett, acting director of the Biomedical Advanced Research and Development Authority within the Office of the Assistant Secretary for Preparedness and Response.

BARDA was established in December 2006 to support the advanced research development and procurement of medical countermeasures against chemical, biological, radiological, and nuclear threats, pandemic influenza, and emerging diseases such as Ebola and Zika.

In 2010, BARDA initiated its Broad Spectrum Antimicrobials program, or the BSA program that I'll refer to, and in 2016, Congress awarded BARDA substantial new funding specifically to combat antibiotic-resistant bacteria. The BSA program was established to support the development of new classes of antibiotics for both biodefense and commercial applications. Under the auspices of the program, we now manage seven public-private partnerships and a pipeline of nine antibacterials.

BARDA's objectives have been to revitalize the antibiotic pipeline, emphasize programs that address the immediate public health threat of multidrug-resistant organisms, and enhance our biodefense capabilities. We have achieved notable success. Our portfolio is quite mature at this point. Five BARDA products are in phase 3 clinical development, and two have completed the pivotal studies required for FDA approval.

Several products show promise for the treatment of infections due to carbapenem-resistant Enterobacteriaceae or CRE, one of the most urgent threats that we face, and a few have shown in vitro activity against bacteria harboring the MCR–1 colistin-resistance gene recently identified in a patient in Pennsylvania. And we are working with our international partners. One of our partnerships has facilitated groundbreaking coordination of funding with the EU's Innovative Medicines Initiative to speed the development of a drug called aztreonam-avibactam.

BARDA also supports the development of improved clinical diagnostics. Without better diagnostics to guide antibiotic therapy, it will be difficult to prevent the emergence of resistant bacteria. BARDA would thus support a diverse set of tools to differentiate viral and bacterial infections, identify resistant infections in the physician's office, and measure antimicrobial susceptibility. BARDA aims to simplify genetic sequencing tools so that they can be used in clinical laboratories for the evaluation of resistant infections.

Finally, BARDA and NIH will offer a $20 million prize for the development of truly novel diagnostics for drug-resistant bacteria. We will be announcing more details about this prize later this year.

To achieve these goals, BARDA employs innovative partnership models. Within HHS, BARDA has pioneered the use of portfolio partnerships that can advance multiple drug candidates simultaneously. BARDA has established these flexible cost-sharing business partnerships using our other transaction authority. We have found that portfolio-based funding reduces risk by allowing for the reallocation of resources across activities and among drug candidates as technical and business risks materialize. Clearly, the early-stage antimicrobial pipeline is too thin. To enrich it, BARDA and NIAID are working together to establish an initiative we are calling the CARB accelerator. Innovation frequently occurs in small biotechnology companies and in academic laboratories with limited resources and expertise to move product candidates forward. In such circumstances, promising early-stage candidates often fail to advance. The accelerator will serve as an incubator for new products and explicitly seeks to improve success rates in the early-stage antibiotic pipeline.

BARDA will launch the accelerator in 2016 and support the program for at least 5 years while NIAID will provide an array of product development support services. Both entities will collaborate in managing the program and its investments through a joint oversight committee. Stay tuned. We certainly will have a lot more to report on this initiative in coming years. In summary, as new forms of antibiotic resistance continue to spread worldwide, the prospect of bacterial strains resistant to all available antibiotics can no longer be ignored. Developing new tools for diagnosing and treating drug-resistant bacteria will be essential to preserving the practice of modern medicine. ASPR and its partners play a critical role in leading the charge against such threats, and we look forward to working with Congress to address the global challenge of antimicrobial resistance. I look forward to addressing your questions. [The prepared statement of Richard J. Hatchett follows:]

WRITTEN TESTIMONY, HOUSE COMMITTEE ON ENERGY AND COMMERCE, SUBCOMMITTEE ON OVERSIGHT AND INVESTIGATIONS, "COMBATTING SUPERBUGS: U.S. PUBLIC HEALTH RESPONSES TO ANTIBIOTIC RESISTANCE"

Statement of Richard J. Hatchett, MD

Acting Deputy Assistant Secretary and Acting BARDA Director Office of the Assistant Secretary for Preparedness and Response U.S. Department of Health and Human Services

For Release on Delivery Expected at 10:00 a.m. June 14, 2016

Good morning, Chairman Murphy, Ranking Member DeGette, and Members of the Committee. Thank you for the opportunity to testify before you today about our government's efforts to combat antimicrobial-resistant bacteria. I am Dr. Richard Hatchett and I serve as the Acting Director of the Biomedical Advanced Research and Development Authority (BARDA), a component of the Office of the Assistant Secretary for Preparedness and Response (ASPR) in the U.S. Department of Health and Human Services (HHS). In this capacity, I also serve as the Acting Deputy Assistant Secretary for Preparedness and Response.

BARDA was established by the Pandemic and All-Hazards Preparedness Act (PAHPA) in December 2006 to support the advanced research, development and procurement of novel and innovative medical countermeasures such as vaccines, antimicrobial drugs, and medical devices (including diagnostics) to address the medical consequences of chemical, biological, radiological, and nuclear (CBRN) agents of terrorism and naturally-occurring and emerging threats, such as the 2009-H1N1 influenza, H7N9 influenza, Ebola, and Zika viruses. Through the Public Health Emergency Medical Countermeasure Enterprise (PHEMCE), ASPR is working on efforts to address the growing threat of antimicrobial resistance, which has expanded in recent years. I appreciate the opportunity to highlight our progress to-date and underscore the significant challenges remaining.

BARDA works with our HHS and other federal agency PHEMCE partners to transition medical countermeasures from early development into advanced development and ultimately to an application for Food and Drug Administration (FDA) regulatory review and approval. Advanced development includes critical steps needed for a product to be ready for use. This involves optimizing manufacturing processes so products can be made in quantity to scale, creating and optimizing assays to assure product integrity, conducting late-stage clinical safety and efficacy studies, and carrying out pivotal animal efficacy studies that are often required for approval. Since 2006, BARDA has managed the advanced development of nearly 150 medical countermeasures for CBRN threats and pandemic influenza. Six of these products have received FDA approval in the last two years alone.

Today, we face a growing threat that has the potential to impact more Americans than an infectious disease outbreak or terrorist attack with a biological agent. Antibiotics and other antimicrobial agents have been effective in treating common infections since their widespread use in the 1940s. They are essential to the practice of modern medicine. The problem of antimicrobial resistance was recognized shortly after the first antibiotics, the sulfonamides, were introduced in the late 1930s, and has always presented concerns. Historically, when an infection demonstrated resistance to one class of antibiotics, clinicians would simply switch to a different class. However, the past two decades have witnessed a significant increase in the spread of organisms resistant to multiple classes of antibiotics. Compounding the problem, the pace of introductions of new antibiotics has slowed considerably largely because there is not a sufficient private sector incentive to invest in antibiotic development.

Gram negative bacteria, in particular, have become increasingly resistant to available antimicrobial drugs, and deadly and extremely difficult-to-treat pathogens such as carbapenem-resistant Enterobacteriaceae (CRE) have become more widespread. Colistin, which has long been considered a drug of last resort due to its serious side effects, is being used increasingly to treat infections with such bacteria. The recent emergence of a plasmid-borne colistin resistance gene, *mcr-1*, thus portends the emergence of truly pan-drug resistant bacteria, for which no current antibiotics are effective. The report three weeks ago of *mcr-1*-carrying *E. coli* in a patient with a urinary tract infection in Pennsylvania starkly highlights the urgent need for proactive action to combat antimicrobial resistance.

Along with our PHEMCE colleagues, BARDA has worked aggressively over the last six years to meet this threat and to re-engage academia and industry in the development of new antibiotics and antiviral drugs. Consistent with an ASPR strategic priority to invest in biodefense products and technologies, which also support everyday applications, BARDA's development of new antibiotics and antiviral drugs is critical to our defense from a naturally occurring threat or terrorist attack. It also has the potential for substantial day-to-day benefit for health care across the U.S. In 2010, BARDA established a Broad Spectrum Antimicrobial (BSA) program to support the development of new classes of antibiotics. Over time we have built out our portfolio and now manage a pipeline of nine products.

BARDA's objectives in establishing the BSA program were to: (1) revitalize the antibacterial pipeline through the support of public-private partnerships targeting novel unprecedented and precedented classes of antibiotics; (2) emphasize programs that address the immediate public health threat of multidrug resistant strains of hospital and community acquired pathogens; (3) provide a biodefense capability to bridge the response between the first clinical case of threat agent infection to when mass dispensing of medical countermeasures is initiated; and (4) enable the evaluation of products for all biothreat pathogens via the development of animal models and tools to support regulatory approval.

We have achieved notable successes in this area. We have built a robust portfolio of broad spectrum antibiotics that possess activity against Gram negative bacteria, which is noteworthy given the limited number and increasing ineffectiveness of antibiotics against these deadly and difficult-to-treat bacteria. Five of our programs have reached Phase 3 clinical development, the last stage before an application for FDA approval, and two products have completed the pivotal registration clinical studies required for approval. One relationship, with Astra Zeneca, is noteworthy in that it reflects a groundbreaking effort to coordinate funding with the EU's Innovative Medicines Initiative (IMI) and thereby speed the development of Astra Zeneca's lead antibiotic, aztreonam/avibactam.

In 2014, the President's initiative on Combating Antibiotic-Resistant Bacteria (CARB) directed BARDA to "develop new and next-generation countermeasures that target antibiotic-resistant bacteria that present a serious or urgent threat to public health" through our advanced development of both drugs and diagnostics.

As outlined in the companion National Strategy and Action Plan for CARB, PHEMCE partners were given clear goals for addressing antimicrobial resistance, objectives for meeting those goals, and milestones to measure progress and success. ASPR, in cooperation with its federal partners, has been engaged in laying out a strategy to address the medical consequences of antimicrobial resistance. Within HHS, we are engaged in strategic advisory boards and working groups dedicated to addressing many of the requirements in the national action plan such as the Interagency Task Force on Antimicrobial Resistance, the Transatlantic Taskforce on Antimicrobial Resistance, and the National Strategy for Research and Development working group.

The recent identification of an *mcr-1*-carrying *E. coli* bacteria in a patient with a urinary tract infection in Pennsylvania highlights the serious health threat posed by drug-resistant bacteria. The nine antibacterial products that BARDA is developing address a range of drug-resistant bacterial threats, including *E.coli*, that were identified as either urgent or serious by the CDC in their 2013 report on the public health crises of antibiotic-resistant infections. A number of these have shown promise for the treatment of infections due to carbapenem-resistant Enterobacteriaceae (CRE). A few of the drugs in BARDA's portfolio have already shown direct *in vitro* activity against bacteria harboring *mcr-1* (so-called "superbugs") and if approved these products could provide some relief from resistance concerns in the future. More broadly, each of the products in the BARDA portfolio is a broad spectrum antibiotic that possesses activity against Gram negative bacteria, which is important because comparatively few antibiotics active against Gram negative bacterial infections are in development and new drugs to treat such infections are desperately needed. BARDA is actively working with industry to pursue innovate approaches and strategies to improve our nation's arsenal against antibiotic resistant organisms.

Combating antibiotic resistance is a public health and national security priority, and BARDA has developed a number of innovative partnership models to support this goal. In recent years, BARDA has pioneered the use of portfolio partnerships, in which companies investigate the advancement of multiple drug candidates at the same time. Funding a portfolio of products, rather than just a single candidate at a time increases the probability of bringing a successful drug to market. BARDA has established these partnerships using the PAHPA-granted Other Transaction Authority (OTA) to create flexible, cost-sharing business partnerships between the government and industry. Portfolio-based funding reduces risk by allowing for the reallocation of resources across activities and among drug candidates if technical or business risks materialize. Since 2013, BARDA has successfully used its OTA twice to establish public-private partnerships. The establishment of the partnership with Astra Zeneca fulfilled a requirement in the CARB National Action Plan to create at least one additional portfolio partnership with a pharmaceutical or biotechnology company by March 2016 to accelerate development of new antibacterial drugs, and several others are currently in negotiations.

BARDA works closely with other government funding agencies, specifically the National Institute of Allergy and Infectious Diseases (NIAID) within the National Institutes of Health, to provide support of product development from early stage discovery through to approval and procurement. While NIAID primarily supports early stage discovery of new drugs, vaccines, and diagnostics and preclinical product development up through Phase 1 clinical studies, programs normally become eligible for BARDA funding once an application to administer an investigational drug to humans is filed with FDA and the project is in Phase 1 development or later. BARDA works closely with our interagency partners, including NIAID and the programs in the Department of Defense, to identify and transition projects within their portfolios to BARDA where we can support the later stage clinical development of products from Phase 1 to approval.

In today's marketplace, innovation often occurs in small biotechnology companies and in academic laboratories that lack the resources and expertise to move candidates along the development path from lead optimization to clinical development. Many promising early stage candidates, without the appropriate resources and support, will be unable to advance to late stage clinical development and approval. To overcome this problem, BARDA and NIAID are working together to establish and operationalize a new initiative we are calling the CARB Accelerator.

The CARB Accelerator will serve as an incubator for new products and explicitly seeks to reinvigorate the early stage antibiotic pipeline and repopulate the late stage clinical development pipeline with new and promising antibiotics. The CARB Accelerator will fund research and development activities to help progress candidate products from the proof-of-concept stage through pre-clinical development. BARDA will provide direct funding to launch the Accelerator in 2016, with a goal to support the program thereafter for five years, while NIAID will provide access to an array of product development support services. NIAID and BARDA will collaborate in managing the program and its investments through a joint oversight committee. BARDA's goal for the Accelerator is to stimulate the early stage development of new antibiotic candidate products and to progress products to the step of filing an FDA Investigational New Drug (IND) application, which is required before drugs can enter human clinical trials. After receiving the FDA's authorization, products can then enter into the clinical development phase and thus become eligible for direct BARDA support. Successful products will emerge from the Accelerator as graduates with resources, networks and a robust product development strategy to support further clinical development.

In 2016, the Accelerator will accept proposals from innovators seeking support for the nonclinical development of their candidate products. BARDA looks forward to engaging the community of researchers and product developers through another novel public private partnership and re-inventing how government is able to partner with industry to address the urgent problem of antimicrobial resistance.

As new forms of antibiotic resistance continue to spread worldwide, the prospect of bacterial strains resistant to all available antibiotics can no longer be ignored; the continued development of new options for treating drug-resistant bacteria will be essential to preserving our ability to effectively combat infections. ASPR and its partners will continue to play a critical role in leading the charge against these threats, both present and future. We look forward to working with Congress to develop new antimicrobial therapies and non-traditional approaches to address the global challenge of antimicrobial resistance. Thank you again and I look forward to your questions.

Mr. MURPHY. Thank you, Doctor.

Now, Dr. Dixon, you are recognized for 5 minutes.

STATEMENT OF DENNIS M. DIXON

Mr. DIXON. Thank you, Mr. Chairman, Ranking Member DeGette—

Mr. MURPHY. Make sure the microphone is on and pulled close to you, OK?

Mr. DIXON. Mr. Chairman, Ranking Member DeGette, and members of the subcommittee, thank you for the opportunity to discuss antibiotic resistance, a serious and growing global health threat. I am Dr. Dennis Dixon, and I serve as chief of the Bacteriology and Mycology Branch in the National Institute of Allergy and Infectious Diseases.

The NIAID is the lead institute at the NIH for research on infectious diseases, including research on antibiotic resistance. NIAID's longstanding research efforts in this area aim to understand the molecular basis of antibiotic resistance, as well as to develop specific and sensitive diagnostics, vaccines to prevent infections, and to partner with pharmaceutical industry companies in the development of novel and improved treatments.

The recent detection in the U.S. of bacteria resistant to colistin, an antibiotic of last resort, reminds us of the urgent challenge of drug resistance and the need to address its underlying causes. Fortunately, this particular bacterial strain was treatable by other antibiotics. However, the threat remains that this type of resistance could emerge in other bacteria resistant to most antibiotics making them untreatable with currently available drugs.

Through our research mission, NIAID plays a critical role in the administration's national strategy and action plan on Combating Antibiotic-Resistant Bacteria or CARB. NIAID's antibiotic resistance research portfolio includes basic research on how bacteria develop resistance and cause disease; translational research to develop diagnostics, therapeutics, and vaccines; and clinical research to evaluate antibacterial products and strategies.

Additional funding provided by Congress in fiscal year 2016 for CARB have been instrumental to our efforts. NIAID-supported basic research provides the foundational knowledge essential for the development of vaccines to prevent antibiotic-resistant infections and therapeutics to treat them.

As part of the CARB initiative, NIAID supports genome sequencing for a national database of antibiotic-resistant bacteria. This database, developed by NIH in collaboration with FDA and CDC, will provide a comprehensive resource for surveillance, epidemiology, and basic research into the mechanisms of antibiotic resistance.

NIAID also facilitates product development by providing non-monetary support to researchers, including genome sequences, access to clinical specimens, drug screening, and animal model testing. These resources provided much-needed support for the field and help to reduce the risk for product developers.

We also provide clinical trials capacity for evaluating new antibacterial products and strategies through our Antibacterial Resistance Leadership Group and other clinical trial networks for clinical research on antibiotic resistance. We are currently in the process of expanding this infrastructure.

NIAID is also pursuing the development of point-of-care diagnostic tests critical to determining which drugs will be effective against given infections and recently awarded funding for research projects that develop rapid, sensitive, and specific diagnostic tools.

NIAID is working to discover better treatments to treat antibiotic-resistant infections by screening new compounds and optimizing the use of existing drugs. NIAID recently awarded funding for research projects to develop nontraditional therapeutics for bacterial infections and funds clinical studies testing new formulations, dosing regimens, or combination therapies of currently licensed drugs such as colistin.

In summary, NIAID is committed to a robust and comprehensive research effort to address antibiotic resistance and is fostering collaborations with partners in academia, industry, and the Federal Government. NIAID will continue support promising research to develop and test new antibiotics as well as methods to prevent the further spread of antibiotic resistance.

Thank you for bringing attention to this important topic. I'd be pleased to answer any questions.

[The prepared statement of Dennis M. Dixon follows:]

DEPARTMENT OF HEALTH AND HUMAN SERVICES NATIONAL INSTITUTES OF HEALTH, THE ROLE OF THE NATIONAL INSTITUTE OF ALLERGY AND INFECTIOUS DISEASES RESEARCH IN COMBATING ANTIBIOTIC RESISTANCE, TESTIMONY BEFORE THE HOUSE ENERGY AND COMMERCE COMMITTEE, SUBCOMMITTEE ON OVERSIGHT AND INVESTIGATIONS

Dennis M. Dixon, PhD
Chief, Bacteriology and Mycology Branch
Division of Microbiology and Infectious Diseases
National Institute of Allergy and Infectious Diseases
June 14, 2016

Mr. Chairman, Ranking Member DeGette, and members of the Committee, thank you for the opportunity to discuss antibiotic resistance, a serious and growing global health threat. The National Institute of Allergy and Infectious Diseases (NIAID) is the lead institute at the National Institutes of Health (NIH) for conducting and supporting research on infectious diseases, including research on antibiotic resistance.

NIAID has a dual mandate to balance a robust research portfolio in established infectious and immunologic diseases with the capacity to respond quickly to newly emerging and reemerging infectious diseases.

Infections resistant to currently available antibiotics are among the most urgent of these emerging threats.

Antibiotic resistance is a multifaceted problem, and multiple approaches are being undertaken to work toward a comprehensive solution. These include bolstering surveillance, diagnostic capacity, hospital infection control, and the prudent use of antibiotics in both humans and animals. Biomedical research also is essential to address the problem. NIAID's longstanding research efforts in this area aim to understand the molecular basis of antibiotic resistance, to develop specific and sensitive diagnostics, to develop vaccines to prevent infections prone to resistance to antibiotics, and, importantly, to partner with the pharmaceutical industry to develop novel and improved interventions.

Antibiotic Resistance

The development and use of antibiotics to treat bacterial infections are among the greatest achievements of modern medicine. However, many of these drugs have become less effective over time as resistance to them has emerged and spread globally. The Centers for Disease Control and Prevention (CDC) estimates that each year in the United States, at least 2 million people become infected with antibiotic-resistant bacteria and at least 23,000 people die as a result of these infections. CDC has identified the most serious antibiotic resistance threats in the United States including carbapenem-resistant Enterobacteraciae (CRE), *Clostridium difficile, Neisseria gonorrohoeae,* and drug-resistant strains of *Staphylococcus aureus* such as methicillin-resistant *S. aureus* (MRSA).

The recent detection in the United States of bacteria resistant to colistin, an antibiotic of last resort, reminds us of the urgent challenge of drug resistance and the need to address its underlying causes. We were fortunate that this particular bacterial strain was treatable by other antibiotics; however, the threat remains that this type of resistance could emerge in other bacteria already resistant to most antibiotics, making them untreatable with currently available drugs.

White House Initiative on Combating Antibiotic-Resistant Bacteria

In 2014, the Administration announced a comprehensive set of new federal actions to combat the rise of antibiotic-resistant bacteria and protect public health, including the National Strategy and National Action Plan for Combating Antibiotic-Resistant Bacteria (CARB). Key participants in the CARB effort include NIH; CDC; the Food and Drug Administration (FDA); the Office of the Assistant Secretary for Preparedness and Response (ASPR), including the Biomedical Advanced Research and Development Authority (BARDA); the United States Department of Agriculture; the Department of Defense; and others.

NIAID plays a critical role in the CARB initiative through our research mission. Additional funding provided by Congress in fiscal year (FY) 2016 has been instrumental to our efforts, particularly in the areas of microbial genome sequencing, diagnostics and drug development, and clinical research, as described below.

In addition, NIAID is participating in the Administration's National Action Plan for Combating Multidrug-Resistant Tuberculosis (MDR TB). The emergence of TB that is resistant to multiple antibiotics is a growing global concern. NIAID is leading research efforts to accelerate basic and applied research and development to combat MDR TB. This research complements other CARB efforts that target this increasingly prevalent drug-resistant infection.

NIAID Research Addressing Antibiotic Resistance

As outlined in the 2014 report entitled, "NIAID's Antibacterial Research Program: Current Status and Future Directions," NIAID supports a robust research portfolio that includes basic research on how bacteria develop resistance and cause disease; translational research to develop diagnostics, therapeutics, and vaccines; and clinical research to evaluate antibacterial products and strategies.

Basic Research and Research Resources to Facilitate Product Development

NIAID intramural scientists and extramural grantees are conducting basic research to understand microbial pathogenesis, including how bacteria colonize our bodies and evade immune defenses. NIAID-supported research also aims to discover mechanisms of antibiotic resistance. Basic research provides the foundational knowledge essential for the development of vaccines to prevent antibiotic-resistant infections and therapeutics to treat them. For example, NIAID researchers discovered that *S. aureus* produces peptide toxins critical for bacterial growth and pathogenesis in a mouse model, suggesting that the toxins themselves may be a target for development of new drugs for this important pathogen that may become resistant to methicillin or other drugs.

High-throughput genome sequencing efforts also are leading to a better understanding of bacterial pathogenesis and how antibiotic resistance develops. NIAID supports genome sequencing for a national genome sequence database of antibiotic-resistant bacteria as part of the CARB initiative. This database, being developed by NIH in collaboration with FDA and CDC, will provide a comprehensive resource for surveillance, epidemiology, and basic research into the mechanisms of antibiotic resistance. NIAID also funds a large-scale sequencing project to understand the genetics of drug resistance in TB. In addition, NIAID scientists and their colleagues have sequenced the complete genome of a drug-resistant strain of *Klebsiella pneumoniae,* which is a significant cause of hospital-acquired infections. This sequencing effort has revealed two different lineages of these bacteria with separate evolutionary histories, providing insight into the multiple pathways through which bacteria become resistant to antibiotics. Additionally, the NIAID-supported Bioinformatics Resource Centers have assembled the genome of the *Escherichia coli* containing the colistin resistance gene, *mcr-1*, isolated from the first patient in the United States (mentioned above), and has rapidly made it available to the research community for further study.

NIAID also provides a comprehensive set of research resources and services designed to reduce the risk to product developers and to help move concepts and candidate products along the research pipeline. These include genome sequencing, quality-controlled research reagents and clinical specimens, drug screening, and animal models to accelerate antibacterial product discovery.

In addition, NIAID is expanding its Antibiotic Resistance Leadership Group (ARLG) and other clinical trial networks, such as the Vaccine and Treatment Evaluation Units, for clinical research on antibiotic resistance. This expansion, facilitated by the CARB initiative and associated funding, will increase the capacity to evaluate new antibacterial products and strategies and move needed countermeasures along the research and development pipeline.

Diagnostics

Rapid, point-of-care diagnostic tests can be important in determining precisely which drugs will be effective against a given infection, thereby reducing the inappropriate use of broad spectrum antibiotics. Currently, broad-spectrum antibiotics that target a wide range of bacteria are often prescribed when a diagnosis is not available prior to starting treatment. The CARB initiative is incentivizing the development of rapid, point-of-care diagnostics to identify antibiotic-resistant bacteria and inform treatment of these infections. NIH and BARDA are collaborating to fund a $20 million diagnostic prize, and NIH held a public consultation and web-based forum to solicit comments on the technical criteria and performance characteristics of diagnostics to be considered for the prize. In addition, NIAID recently awarded more than $11 million in first-year funding for research projects supporting enhanced diagnostics to detect antibiotic-resistant bacteria. The grantees in this program will work to develop rapid, sensitive, and specific diagnostic tools that do not rely on the time-consuming process of growing the bacteria, which is how bacterial infections are often diagnosed currently.

NIAID also supports clinical studies through its ARLG to improve diagnosis of drug-resistant infections. The ARLG is developing a study to test procalcitonin, a biomarker that could help clinicians distinguish between bacterial and viral lower respiratory tract infections. Such a biomarker could help to reduce inappropriate use of antibiotics for viral infections, an important driver of the development of antibiotic resistance. Other ongoing ARLG studies are comparing the performance of existing diagnostics and validating new tools including platform diagnostics to rapidly identify antibiotic-resistant bacteria.

NIAID also has supported the development and validation of a test that can rapidly identify TB and simultaneously detect resistance to rifampicin, an antibiotic commonly used to treat TB. This test, and next-generation versions of the test, currently are being implemented in developing countries to diagnose TB, including drug-resistant TB.

Therapeutics

NIAID is screening new compounds and repurposing existing drugs to provide better options to treat antibiotic-resistant infections. In addition, many novel approaches are being explored, including monoclonal antibodies, bacteriophages, and strategies targeting the microbiome or the host immune system. NIAID recently awarded approximately $5 million in funding for 24 research projects seeking to develop non-traditional therapeutics for bacterial infections. These awards are investigating novel therapies such as bacteriophages, probiotics, and nanoparticles to treat or prevent infections. NIAID intramural researchers also are exploring a novel treatment for TB called host-directed immunotherapy. Rather than targeting the bacteria directly, this approach involves manipulating the body's response to TB, using a regimen that includes zileuton, a clinically approved drug for asthma, to target components of the immune response. In addition, NIAID-supported scientists recently identified the drug teixobactin using an innovative iChip platform that allows researchers to screen natural products from bacteria that live in soil. This drug has a novel mechanism of action and has shown promise against several antibiotic-resistant microbes. Although teixobactin is still under development, potentially it could be a new tool to treat drug-resistant bacteria.

Optimizing the use of existing drugs also can help limit the development of antibiotic resistance. To this end, NIAID funds clinical studies testing new formulations, dosing regimens, or combination therapies of currently approved drugs such as colistin. NIAID-supported researchers recently found evidence that two off-patent antibiotics, clindamycin and trimethoprim/sulfamethoxazole, work equally well against bacterial skin infections caused by MRSA, indicating that these infections can be treated successfully and inexpensively with either therapy. In addition, an ongoing NIAID-supported Phase IV clinical trial is comparing different combinations of existing antibiotics for treatment of gonorrhea, a disease for which drug resistance is a growing concern.

Vaccines

Developing vaccines to prevent infectious diseases can help prevent the inappropriate use of antibiotics and the development of antibiotic resistance. NIAID is developing vaccines to prevent infections for which treatment options are jeopardized by the emergence of drug resistance, notably staphylococcal infections and gonorrhea. Vaccines for viral infections such as influenza also may help reduce the use of antibiotics, which are often used inappropriately to treat viral infections, or appropriately used to treat bacterial infections that sometimes develop following viral infections. NIAID is working to develop vaccines for influenza, including several promising universal influenza vaccine candidates. A universal influenza vaccine could reduce or eliminate the need for annual seasonal influenza vaccines and provide long-lasting protection against multiple strains of influenza, including seasonal and pandemic influenza. Such a vaccine could lessen the burden of influenza and associated secondary bacterial infections, thereby reducing the need for antibiotics.

Conclusion

NIAID is committed to a robust and comprehensive research effort to address antibiotic resistance and is fostering collaborations with partners in academia, industry, and the federal government. NIAID will continue to support promising research to develop and test new antibiotics as well as methods to help prevent the further spread of antibiotic resistance.

Mr. MURPHY. I thank all the panel for their comments here. I am going to recognize myself for 5 minutes of questions.

On September 20 of 2000, then CDC director Dr. Jeffrey Koplan testified before Senate Appropriations Subcommittee on Labor, HHS, and Education on the emerging national and global problems of antimicrobial resistance and the response by CDC. And at that hearing the CDC unveiled its plan that targeted drug resistance. In particular, Dr. Koplan testified that a key part of this plan was developing a national campaign to improve physician prescribing practices and educate parents and patients about the proper use of antibiotics.

Dr. Bell, despite the CDC's campaign as well as the efforts of others, the overprescribing of antibiotics still exists. So why did the overprescribing and overuse of antibiotics continue even with the CDC's campaign?

Dr. BELL. Thank you very much, Mr. Chairman.

The campaign that I think you're referring to is a campaign which is called Get Smart, and this is something that we've been working on for quite some time. But actually, we've had quite a bit of impact. Actually, we've been able to show decline. The Get Smart campaign was really focused on outpatient prescribing. And while we are very concerned about the continued overuse of antibiotics in outpatients, we've actually seen considerable improvements over the years and they've been able to show an impact of the Get Smart program.

Antibiotic overprescribing and stewardship is a complex issue, and we really need to attack it on many different fronts. The Get Smart and outpatient facilities is just one piece of the puzzle, and we've been really redoubling our efforts in the area of stewardship over the last few years releasing guidelines for stewardship programs and essential elements for hospitals and for long-term care facilities.

Mr. MURPHY. Do you see trends changing, that it is getting through to physicians because——

Dr. BELL. Well, we now know that there are about 40 percent of hospitals in the United States that have all the elements of a stewardship program, and we've set a target of 100 percent of inpatient facilities by 2020. There are a lot of positive developments. I think——

Mr. MURPHY. But go back to that number of the huge——

Dr. BELL. Yes.

Mr. MURPHY [continuing]. Amount of prescriptions written every year——

Dr. BELL. Yes.

Mr. MURPHY [continuing]. That shouldn't be. So that doesn't sound like it is working to me. I just want to know what else—we are here to help you.

Dr. BELL. Yes.

Mr. MURPHY. We are all on the same team here. I know the White House Action Plan at Combating Antibiotic-Resistant Bacteria——

Dr. BELL. Yes.

Mr. MURPHY [continuing]. It is there. How is that going to be different? What do we see that is really going to get through to people this time?

Dr. BELL. Well, we really I think are just over the last few years really attacking this on lots of different fronts. As I said, the initial Get Smart program was really an educational campaign focused in one area, and now we have many different modalities. We've been collaborating very effectively also with CMS, with the American Hospital Association. CMS will be developing conditions of participation both for hospitals and long-term care that will require——

Mr. MURPHY. Will this also—excuse me.

Dr. BELL. Excuse me.

Mr. MURPHY. I have to jump in because we are short on time.

Dr. BELL. Oh, sorry.

Mr. MURPHY. Will this also include—and I ask the panelists. Will this also include information for patients?

Dr. BELL. Yes——

Mr. MURPHY. I mean, let's face it. People go to the doctor and say I want a prescription. He says, well, you have got a virus, it is not going to work; I want it anyways. And doctors now especially because they get rated on——

Dr. BELL. Yes.

Mr. MURPHY [continuing]. Were you treated satisfactorily and hospitals then get dinged. They don't get paid as much.

Dr. BELL. Yes.

Mr. MURPHY. And I have heard time and time again hospitals say, look, we are afraid of these ratings and so they will show up in Web sites, everything else. So what are we doing to combat that?

Dr. BELL. I appreciate your point. And we have been developing a lot of educational campaigns focused on patients. And I think you're right. Our objective would be for patients, when they go to the doctor and the doctor wants to prescribe an antibiotic, the patient says why do I need this antibiotic? Please explain that to me. And we do really need a sea change, and we're certainly working in lots of different ways to educate the patient. The patient really is pivotal in terms of solving this problem.

Mr. MURPHY. Dr. Woodcock, it is always good to see you. Why haven't we developed rapid diagnostic testing for bacterial infections yet? And can you tell us about some of the commercial and technical impediments here?

Dr. WOODCOCK. Well, it's just harder than it sounds. There are a few. We have strep test, and that's great, OK, so there's something where people can differentiate quickly and put people at ease perhaps that they don't have—don't need an antibiotic, an outpatient setting, right? The technology is advancing, as I said in my oral statement. The Center for Devices is seeing technologies come along in the development pipeline so we expect over the next several years we will see more rapid diagnostic tests in a variety of settings.

It's urgent that we try to find some that differentiate bacterial and virus infections. People are working on expression profiles that you can do a blood test and make that determination, and then you could reassure the patient you don't have a bacterial infection. And that would be very helpful.

Mr. MURPHY. Thank you. I see I am out of time so I will recognize Ms. DeGette for 5 minutes.

Ms. DEGETTE. Thank you, Mr. Chairman.

As I said in my opening, I want to talk about how we can incentivize development of new antibiotics, and as we heard, earlier this year BARDA and NIAID collaborated to establish the Combating Antibiotic-Resistant Bacteria, or CARB, by a pharmaceutical accelerator. So, Dr. Hatchett, I am wondering if you can tell me very briefly why this program is important.

Dr. HATCHETT. Well, thank you, Ms. DeGette for the question. As I said in my opening testimony, BARDA is very concerned. We do currently have a mature portfolio of antibiotics that are close to achieving licensure, but we are concerned that the upstream pipeline is very weak——

Ms. DEGETTE. Right.

Dr. HATCHETT [continuing]. And that reflects the progressive disinvestment over several decades by biopharmaceutical firms not viewing antibiotics as providing sufficient return on investment.

The goal of the biopharmaceutical accelerator will be in collaboration with our partners at NIAID, and we've worked very closely with them to structure the accelerator to support that early development, bring resources and capabilities to early-stage innovators and help them move rapidly through the development process and advance those products hopefully to a point where they can transition to direct support for it.

Ms. DEGETTE. Thank you. And, Dr. Woodcock, I want to follow up on this because when you testified before this committee in 2014 about 21st Century Cures, you talked about this issue of lack of commercial incentives for drug developers and that being a reason why we don't have new investigational drugs. Can you quickly update us on what the situation is since 2014 with this? Do we have some promising drugs in the pipeline? Are we fixing some of these commercial issues?

Dr. WOODCOCK. There are some promising drugs in the pipeline, but the pipeline is still very fragile. And what we need is not just a few superstars here. We need a full panoply of investment in the research, the basic science, the drug discovery and drug development that lasts, as Dr. Hatchett just said, over decades because we don't just need a few antimicrobials, we need a whole continuing——

Ms. DEGETTE. Range. Range.

Dr. WOODCOCK [continuing]. Platform and range of them for a wide variety of diseases, and we're just not seeing that, not all of the ones under development are going to succeed. And that's true of all drug development.

Ms. DEGETTE. So Chairman Murphy and I got a letter yesterday from the Infectious Disease Society of America, and that organization said in their letter that the administration promised to release a report and recommendations on economic incentives for antibiotics, but the reported recommendations have not been released. I am wondering what the status of that report is and when it will be released.

Dr. WOODCOCK. I don't know. It's not under my purview, but we can get back to you.

Ms. DEGETTE. I wish you would, thanks, because maybe that can help illuminate—yes, Dr. Hatchett?

Dr. HATCHETT. I just want to mention that Secretary Burwell did ask the Presidential Advisory Committee on Combating Antibiotic-Resistant Bacteria to look explicitly at the issue of necessary eco-nomic incentives for the development of antibiotics. That letter was sent to the committee at the end of March, so they are actively undertaking that review.

Ms. DEGETTE. All right. Does anybody here know what this report is and when we are going to get it?

Dr. HATCHETT. The report was an internal report at the White House, and I don't know the status of it. You'll have to address the question to them.

Ms. DEGETTE. Dr. Dixon, can you talk to me about the clinical trials that are going on on strategies for using existing drugs more effectively?

Mr. DIXON. Absolutely. Thank you for the question. And it is important not only to develop new drugs but to optimize the ones we have left. And in fact, we have a clinical trial underway with colistin because we don't know how long it will be before we run out of use of colistin. But knowing how to use it wisely——

Ms. DEGETTE. Right.

Mr. DIXON [continuing]. Will reduce the risk of emergence of resistance.

Ms. DEGETTE. OK.

Mr. DIXON. So we have looked at how the body metabolizes colistin—in other words, the pharmacokinetics—to maximize the presence of drug in the blood and maximize the activity of the drug and minimize the emergence of resistance. And so we have completed a study there that helped to inform better dosing of colistin——

Ms. DEGETTE. Right.

Mr. DIXON [continuing]. And we're now looking at colistin alone versus colistin in combination with another drug for carbapenemresistant Enterobacteriaceae, or CRE, and other resistant pathogens.

Ms. DEGETTE. Thank you.

Mr. DIXON. And so——

Ms. DEGETTE. Thank you very much. I want to ask Dr. Bell a quick question about why improved diagnostics are an important part of addressing the threat of antibiotics resistance.

Dr. BELL. Yes, thank you. As we've been talking, the issue of stewardship is so pivotal to addressing this whole issue of antibiotic resistance, and improved diagnostics, having the capability to be able to differentiate between a viral infection and a bacterial infection right there when the patient is sitting in front of you would be a really important tool that would help us in all of our stewardship efforts. So we really do hope that diagnostics that can at least perform that basic function will be something that'll be available——

Ms. DEGETTE. Thank you.

Dr. BELL [continuing]. Relatively soon.

Ms. DEGETTE. Thank you so much, Mr. Chairman. I yield back.

Mr. MCKINLEY [presiding]. Thank you. I recognize myself for 5 minutes.

This discussion reminds me a little bit of some of the environmental issues we have been dealing with over the last 5 years. The fact that we only represent such a small part, 5 percent of the population of the world, yet we consume so much of our energy, and what Gina McCarthy from the EPA there other day said that it makes no difference what we do; it is what happens around the world is what is really going to affect the environment, so notwithstanding all the issues that we have done in rules and regulations.

So my question goes back to my opening statement, was when with 95 percent of the rest of the world perhaps overprescribing antibiotics and we don't know what is happening in some other nations around the world. We know 20 nations at least and Mexico are over-the-counter for antibiotics. This global issue, what should we be doing? We can solve it here, but if indeed the patients, people are coming in that have drug resistance to our antibiotics, what do we do? So if the four of you could just give me a sense. This fight is a global fight, not a United States. Please, Dr. Bell.

Dr. BELL. Thank you. Yes, you're absolutely right. This is definitely a global fight. These microbes move around the world, and we have to look at this from a global perspective.

So we are working in a number of different areas. So, first of all, we do have a number of really excellent collaborations with the Europeans and the Transatlantic Task Force for Antimicrobial Resistance, which allows us to share information very quickly, develop common standards, and also pool our resources to help other countries around the world. That's a very effective collaboration.

The World Health Organization is actually getting serious about antibiotic resistance. The World Health Assembly has passed several resolutions with all nations basically committing to do something about antibiotic use. So we have to do everything we can to support them——

Mr. MCKINLEY. But even with our education——

Dr. BELL [continuing]. And we are——

Mr. MCKINLEY [continuing]. We are still at——

Dr. BELL. That's right. It's a hard problem. And we haven't solved it here——

Mr. MCKINLEY [continuing]. Forty percent, so even with our education level here——

Dr. BELL. That's right. It's a hard problem.

Mr. MCKINLEY [continuing]. I am troubled about other areas, emerging nations——

Dr. BELL. Yes.

Mr. MCKINLEY [continuing]. What is happening. So please continue.

Dr. BELL. I think——

Mr. MCKINLEY. I am not sure I am buying into all the educational part of it. I am trying to figure out what the other solution is.

Dr. BELL. Yes. No, I think there are many parts of this. It can't just be education, you're absolutely right. There have to be national policies and there has to be national legislation that supports this sort of antibiotic stewardship around the world. And as I say, I think that there is a fair amount of interest in that.

We've been working also a lot on improving detection so that at least we know what's emerging where, and this is something that——

Mr. MCKINLEY. Let——

Dr. BELL. Go ahead. I'm sorry.

Mr. MCKINLEY. Let's go to that point.

Dr. BELL. Yes.

Mr. MCKINLEY. Are there nations that are more prone to have problems that we need to deal with——

Dr. BELL. Well——

Mr. MCKINLEY [continuing]. Around the world? Because we are not all equal so——

Dr. BELL. No, we're not all equal, and there certainly are some countries that do have already a significant problem with, for example, bacteria that are resistant to all antibiotics. Colistin that we've been talking a lot about here is actually used quite a bit more to treat human infections in other parts of the world, in India, for example, because there are unfortunately a large number of patients that show up in intensive care units, for example, that are infected with bacteria that are resistant to all other antibiotics.

So this is a larger problem in some other parts of the world. There are some hot spots. And this issue of detection and focusing on the hot spots and knowing where the hot spots are is something that we've been working on.

The Global Health Security Initiative includes antibiotic resistance as one of the areas that the countries have agreed to focus on so——

Mr. MCKINLEY. Should we be spending more of our monies overseas to try to work those problems out over there?

Dr. BELL. Well, I think that we—it would be not wise for us not to focus on our local problem here. If we don't know what's going on here, we're not going to be able to protect Americans. We have to be doing both because we have to be both strengthening our ability to prevent and detect and respond in the United States while at the same time supporting global efforts.

Mr. MCKINLEY. OK. So two quick questions. One, Colistin, is that a reasonably marketable, affordable drug or is this something——

Dr. BELL. No, colistin is a very old antibiotic. It's been around for a long time. As a physician, I hate to prescribe colistin to patients. And actually in any country doctors really don't want to use it. It has a lot of potential side effects. It has kidney toxicity, it has brain toxicity. So it really is a last-resort antibiotic.

Mr. MCKINLEY. I am afraid my time is expired, but thank you very much for your responses.

If I could, next line of questioning from Mr. Pallone, 5 minutes.

Mr. PALLONE. Thank you, Mr. Chairman.

I wanted to ask each of you a question about investment in fighting antibiotic resistance, so I am going to go through quickly so I can get to each of you.

In the last few years, we have seen increased investment in this fight. The President released the National Action Plan for Combating Antibiotic-Resistant Bacteria in March of last year with ambitious goals for many Federal departments and agencies. And I am encouraged by the over $830 million we directed toward this effort in fiscal year 2016. And the President, as you know, has requested $1.1 billion for fiscal year 2017 for antibiotic resistance.

So let me start with Dr. Bell. In your prepared testimony, you noted that CDC has received $160 million in fiscal year 2016 to implement the National Action Plan noting that, "These resources are transforming how our nation tackles and slows antibiotic resistance comprehensively, efficiently, and systematically."

So just tell me, how will the increased investment requested for the next fiscal year build on this capacity, and why is sustained investment critically important to our success in combating antibiotic resistance?

Dr. BELL. Thank you. So antibiotic resistance is a problem in communities, so a lot of our funding in 2016, for which we are very grateful to Congress, is being used to fund states and large cities to build their capacity to detect and to respond and to prevent. And so the additional funding in 2017 will go towards being able to go even further in that regard. So we'll be able to fund more states to have the kind of antibiotic-resistant prevention programs. We'll be able to strengthen our laboratory networks so that we can look for more types of antibiotic-resistant organisms. And in general, we'll be able to take the kind of investments that we're making in 2016 to more states and to a higher level.

Mr. PALLONE. Thank you. Dr. Dixon, in your testimony you noted that the additional funding for fiscal year 2016 has been instrumental to NIH efforts. Again, why is sustained investment in fighting antibiotic resistance critical to your agency?

Mr. DIXON. Thank you for the question. We appreciate the additional funding, and we're applying that to such key resources as the National Sequence Database that will inform the sequences and tracking genes like the MCR–1 and others; and diagnosis and going particularly after the problem of diagnostics, which could certainly help to reduce the overprescribing whether or not you have a bacterial infection or a viral infection; advancing basic translational and clinical research overall, fundamentally addresses the problems that can help to identify new candidate drugs, new candidate vaccines, and new candidate diagnostics. And to close up, just to say that we are contributing with BARDA the $20 million diagnostics prize to try and draw key manufacturers into the space.

Mr. PALLONE. Well, thank you. So, Dr. Hatchett, you work with industry to bring new antibiotics to market. How is sustained investment important to your work, and what message does that send to your private partners? The same question.

Dr. HATCHETT. Yes, I was going to say. Let me address the question of the value of the funding——

Mr. PALLONE. Sure. Sure.

Dr. HATCHETT [continuing]. For BARDA. As I mentioned, we established our Broad Spectrum Antimicrobials program in 2010. We did that with funding that was provided by Congress for biodefense purposes. And so we were very diligent in making sure that every antibiotic that we were developing had a biodefense application. For better or worse, what that meant was that we had to develop broad spectrum antimicrobials, which are part of the problem that has contributed to the spread of resistance because of the overuse of broad spectrum antimicrobials.

The funding that Congress provided in 2016 is specifically dedicated to the problem of antimicrobial resistance, and it allows us to think in much more nuanced ways about how we can build up an armamentarium to treat patients who have antibiotic-resistant bacteria so we can make investments in vaccines, we can make investments in monoclonal antibodies and alternative therapies. So strategically, it's very important.

In 2016, we are making a number of investments. You've heard about the CARB accelerator. That's a major investment this year. You've heard about the diagnostics prize. We're also going to be expanding our portfolio partnership this year. All of those efforts are going to take years to mature and to convince our industry partners to enter into partnerships with us in an area that they have traditionally viewed as not providing an adequate return on investment. They have to see that sustained commitment of funding over many, many years and recognize that the government is a reliable partner.

Mr. PALLONE. Thank you. I am out of time for you, Dr. Woodcock. Sorry. Thank you, Mr. Chairman.

Mr. MCKINLEY. Thank you. And next, Dr. Burgess from Texas for 5 minutes.

Mr. BURGESS. Thank you, Mr. Chairman.

Probably of all the members on this committee I have prescribed the most doses of antibiotics. Dr. Bucshon, who was here, is actually significantly younger than I am, so I think I probably have seniority on his as far as the doses of antibiotics.

And it is worth considering just a couple of historic notes as we talk about this, 1929, Sir Alexander Fleming noticed the zone of inhibition around a mole that had grown on his agar plate, and that led to the development of what we now know as penicillin. But it was really 15 years later when the Pfizer Corporation figured out how to manufacture this on a large scale so literally making this wonder drug, which previously was kind of a parlor trick, making it available to the masses and having it available to treat our soldiers on the days that they stormed the beaches at Normandy, thus resulting in significantly lower loss of life and limb from those soldiers who took that beach on that heroic day.

My grandfather practiced obstetrics. He was an academic physician. He unfortunately died in 1940 so he practiced obstetrics in the pre-antibiotic days, and I always kind of look at that with kind of wonder and amazement. Here was an individual that chose a profession before the development of good anesthesia, blood banks, and most of all, antibiotics because anyone who has practiced obstetrics knows that without the ability to prescribe good antibiotics and antibiotics that work well, our ability to fix problems has become severely limited. So we want the antibiotics that we have available to continue to work, and we want them to work well.

The JAMA article that was referenced is of interest, and I thank the staff for pulling it for me so quickly. And just glancing through it, it seems like pharyngitis, otitis media, and sinusitis are the illnesses or the diagnoses that sort of achieve the lion's share of the antibiotic prescriptions. And easy to be critical of the doctors and practitioners who are prescribing those antibiotics, but let us not forget that 130 years before penicillin was discovered, the father of our country succumbed to an illness that was treatable by penicillin, a complication of pharyngitis, tonsillitis, that led to a peritonsillar abscess and ultimately took his life. So bacterial diseases are interwoven within our country's history for the last, what, 85 years, have become part of the story of American medicine.

Dr. Dixon, let me just ask you because this is such an unusual situation with this woman that had the infection. She had an Enterobacter infection? Is that correct? Do I understand the story correctly?

Mr. DIXON. I think you do. It might be better of Dr. Bell, who followed that case and did the case study, could talk about the clinical workup of that individual.

Mr. BURGESS. So how in the world did you find this plasmid or this gene that was so unusual? Was there something about this patient's presentation that said we better look for this rare gene?

Dr. BELL. No. Actually, this story really begins with—it's sort of a case study in how beefing up surveillance can really make a big difference in terms of prevention and response. So since we first heard about this MCR–1 identification in China, all of us in the Federal partners, including the Department of Defense, have been looking for evidence of this gene. And so the Department of Defense has a surveillance system where patients with bacterial infections or isolates within their system that have certain characteristics—in this case, show resistance to some of the extended spectrum beta lactamases—are shuttled eventually to the Walter Reed where they fully characterize these unusual isolates. And it was in the context of that that this was actually—that this was identified.

Mr. BURGESS. But there was nothing in the patient's clinical course or her diagnostic presentation that said, oh, boy, we better be looking for this one needle in the haystack?

Dr. BELL. No, that's right, there wasn't. And actually, just to clarify that this bacteria actually is not a pan-resistant bacteria. So the bacteria that infected this patient is actually treatable by some antibiotics, and the patient herself is actually fine. Indeed, in our field investigation and follow-up we've been able to verify that she no longer has that bacteria in her urine.

Mr. BURGESS. Well, that is good news, but I guess the other question that comes up is how do you know this hasn't occurred on other occasions——

Dr. BELL. Absolutely.

Mr. BURGESS [continuing]. In all of the vast number——

Dr. BELL. We——

Mr. BURGESS [continuing]. Of specimens——

Dr. BELL. Right.

Mr. BURGESS [continuing]. That aren't subject to that type of scrutiny?

Dr. BELL. We do not know, and I think that's to the point about what are we doing with the investments that are available to us because of Congress' appropriation is that we are going to be—and this is one of the things I think we've been concerned about for a long time, that, for example, the CDC being the only lab in the country that can actually look for some of these unusual types of resistance, that's not a good state of affairs. And with the additional investments, we're going to be able to have a much more robust system for systematically looking for these and for looking for other and new, more emerging forms of resistance. In terms of responding to this finding, that's really the first thing that we want to do. We need to figure out how bit a problem this is——

Mr. BURGESS. Yes.

Dr. BELL [continuing]. As you say, Congressman.

Mr. BURGESS. Right. Thank you, Mr. Chairman. I will yield back.

Maybe we will have time for a second round.

Mr. MCKINLEY. Our next is Congressman Green from Texas for 5 minutes.

Mr. GREEN. Thank you, Mr. Chairman.

Antibiotic resistance is a significant public health challenge, and we need think creatively to address it. And I want to hear from our witnesses today about the regulatory and scientific challenge in developing new antibiotics, especially focused on what we can do to tackle them.

Dr. Hatchett, what are some of the challenges facing manufacturers in the antibiotics market?

Dr. HATCHETT. Thank you for the question, Congressman Green.

Certainly there are scientific challenges, and I may turn to my colleague Dr. Dixon to let him address the scientific challenges. I'm just a dumb oncologist so this is not my field, but I do understand the medicinal chemistry particularly for gram-negative bacteria, which are some of the bacteria that we are most concerned about is particularly challenging.

From where I'm sitting, I think the biggest challenges relate to the economic incentives in fact to convince companies to make the long-term investment to do the development. But that wasn't your question.

Mr. GREEN. But what I was trying to get at is that we have both scientific challenges but we also have financial challenges because vaccines are not the next miracle drug. In some cases it could be.

But, Dr. Woodcock, would you like to add anything on the issue between——

Dr. WOODCOCK. Well, I do believe that——

Mr. GREEN [continuing]. The economics and the scientific——

Dr. WOODCOCK. I've talked to some of the large pharmaceutical companies. In the past they have run very broad scientific discovery programs and discovery means they're looking for antimicrobials, OK, they're trying to find candidates. And these programs have failed. So down at the science level, this is hard, and I think this is where NIAID's investment in the science of determining what are the bacteria like and how do they generate resistance and what is very important to advance that science because this is a hard area to develop drugs in.

Mr. GREEN. Thank you. Dr. Hatchett, these next questions are for you. Is BARDA exploring any new ways to incentivize development of these new antibiotics and other therapeutics?

Dr. HATCHETT. Yes, sir, we are. I mentioned our use of our other transaction authority that Congress granted to us in the Pandemic and All-Hazards Preparedness Act. And that allows us to work with companies, particularly large companies that have multiple products in their pipeline. And so we can invest in their entire pipeline as opposed to making a specific investment in a single product.

That approach has attracted a great deal of attention from large pharmaceutical partners. We have two of those portfolio partnerships now, one with GSK, one with AstraZeneca, a number of others are actually in negotiations. The companies find it very attracive because it allows them to keep a focus on the development of anti-infectives in a way that makes economic sense to them. So that's very important.

There is an emerging consensus about economic incentives, that it's going to require both a combination of traditional what we call push incentives or investments in R&D, R&D contracts, as well as pull incentives, the market-entry incentives. If you make it to market, there's a guaranteed market, a guaranteed return on investment, potentially a prize.

Mr. GREEN. OK. That brings us to the GAIN Act from last Congress and the ADAPT Act in this Congress, both individually but also as part of the 21st Century Cures. Envision a scenario where more adaptive clinical trials may be used to help drug developers seeking to create the next antibiotic drug effective against drug-resistant bacteria. Dr. Woodcock, can you tell me your thoughts on how the pathway laid out in ADAPT might benefit drug companies in the pursuit of the new novel antibiotics?

Dr. WOODCOCK. Certainly. Well, after a candidate is discovered and it's for a resistant organism, then you have another scientific problem of how do you find these people who are infected who are maybe scattered around and then test the drug in them when they're critically ill and they need to be treated right away. And so doing these clinical trials, even small clinical programs are extremely challenging. And of course we want the drug to be used in the most resistant organisms, so we want it to be tested there to see if it works. But those are hard to find. And so you have this big problem.

And we are interested in having a limited development program where you'd really have a great deal more uncertainty, but you wouldn't use the drug for sinusitis and otitis media and so forth. This would be signaled that there was only a limited amount of information, but the drug could be used in these desperate situations. And that pathway, we think, would be a reasonable pathway to make the drug available but with a stewardship signal.

Mr. GREEN. OK. Thank you, Mr. Chairman. I know I am out of time, but, Dr. Burgess, I have sinus troubles and every once in a while I get an infection. I will not call you for a prescription.

Mr. BURGESS. Flonase, bubba.

Mr. GREEN. I already take everything else.

Mr. MCKINLEY. All right. The chair recognizes Mr. Collins from New York for 5 minutes.

Mr. COLLINS. Thank you, Mr. Chairman. I want to thank the witnesses as well. It has always been a concern to many of us, but maybe to answer a question that I know goes through a lot of families when our kids say I have got an infection, I need an antibiotic.

And many moms and dads say, well, no, if you take these antibiotics, you will become antibiotic resistant, and some day when you are older and you really need one, it is not going to work.

Contrasted to what we are talking about today is it is not a person who becomes antibiotic resistant, which is kind of this old wives' tale versus there are bugs that are figuring out a way to become antibiotic resistant. And a person getting one of those bugs, even if they have never taken an antibiotic in their life, has the same risk. Would that be a fair——

Dr. WOODCOCK. Yes.

Mr. COLLINS [continuing]. Summary?

Dr. WOODCOCK. Yes, Congressman. If I could use that——

Mr. COLLINS. Yes.

Dr. WOODCOCK [continuing]. Myself, it would be helpful. That's—no, that's absolutely right. And that's part of the point about how antibiotic resistance is really a problem for everybody because it's not an individual that develops resistance. It's the bacteria. And these bacteria are constantly replicating and they share genetic material and they spread around the community. So this is exactly—your explanation is exactly correct. And I've often thought we need a better term than resistance because people think that applies to them, and actually——

Mr. COLLINS. As a person.

Dr. WOODCOCK [continuing]. It doesn't. It applies to the microbe itself. So again, we've oftentimes thought, you know, perhaps we need a better way to explain this to break that old idea——

Mr. COLLINS. Well, sure. I mean——

Dr. WOODCOCK [continuing]. That it's a person.

Mr. COLLINS [continuing]. You can actually have in a family someone who should take an antibiotic, but because of this misconception——

Dr. WOODCOCK. Yes.

Mr. COLLINS [continuing]. The parents are saying, no, no, no, no.

Dr. WOODCOCK. Yes.

Mr. COLLINS. We are going to fight this for a few days, fight this for a week——

Dr. WOODCOCK. Yes.

Mr. COLLINS [continuing]. Because we don't want you to become antibiotic resistant.

Dr. WOODCOCK. Yes, you're absolutely right. I mean, antibiotics, as we've heard are—you know, they're a treasure and they're incredibly important and we need to—when we talk about stewardship, we don't mean don't use antibiotics. We mean use the right antibiotic at the right dose for the right duration for the right indication. And it is a very important part of our message.

Mr. COLLINS. Well, I think that is. That is why these hearings can be important.

Now, I have got a question for Dr. Dixon. I understand at Washington State University—and maybe this will translate to Dr. Woodcock—but I understand they have like an e-bandaid, an electromagnetic thing that they are going to put on someone, they pass the electric current through, they have now determined that is producing hydrogen peroxide. And if they control this, they have gotten some interesting results as a way to treat bacterial infections and wounds without antibiotics. Are you familiar with some of that, and do you have a comment on—I know it sounds a little scifiish, but it actually seems to have some possibilities.

Mr. DIXON. I understand the question. I'm not familiar with that particular example. It is entirely consistent, though, with our approach of looking at things other than traditional drugs, so we're looking at exploiting the host immunity a better way. We're exploiting things like bacteriophage. We're exploring things like microbiota using microbial ecology to outcompete and prevent infections and other alternative approaches that don't provoke the emergence of resistance.

Mr. COLLINS. Yes, I think it makes a lot of sense sometimes to——

Mr. DIXON. Absolutely.

Mr. COLLINS [continuing]. Actually think out of the box, and if you had a chance, you should maybe look into this. It was Washington State University last November, about 6 months ago. The results sounded very encouraging because instead of by guess and by golly, they finally figured out why electric currents sometimes work and sometimes didn't, and it was controlling the process and it all came down to the hydrogen peroxide at a level of consistency that would say, on Shark Tank, this one might get funded. Who knows?

Mr. DIXON. Absolutely. A lot of exciting things going on out there.

Mr. COLLINS. So question for you, Dr. Woodcock. Let's say something like this happens. Now, this is an electrical therapeutic if you will. It is not a "drug." What role would the FDA have, and would you have to run clinical trials like it would be a drug or is there something else when it is almost like a device or something instead of a drug?

Dr. WOODCOCK. Well, there has to be a determination first whether it's a drug or device, and we have a process that's been set up by Congress, an office that sorts that out and what has to be done to study it. It depends on whether it's a device, a medical device or a drug, and those are appropriate to the two different kinds of——

Mr. COLLINS. Being one or the other, it would still have to go through your agency for approval——

Dr. WOODCOCK. Yes.

Mr. COLLINS [continuing]. Final signoff if you will.

Dr. WOODCOCK. These type of therapeutics, whether they're devices or drugs, are regulated by the FDA.

Mr. COLLINS. OK. And no idea what this one would be, but I think all of us are encouraged there may be in this high-tech world we live in now something other than antibiotics. At the same time, we have got to keep searching, and it is kind of an all-of-the-above search if you will.

Thank you for your testimony. My time is expired and I yield back, Mr. Chairman.

Mr. MCKINLEY. Thank you. And I now recognize for the next round of questions Ms. Castor from Florida for 5 minutes.

Ms. CASTOR. Thank you, Mr. Chairman. Thank you all for your expert testimony today.

I want to return to the discussion of how we tackle this internationally, as Mr. McKinley focused on, because no single country like the U.S. can do this alone. We are so interconnected now. People travel. You said microbes travel, animals travel, food travels.

In July of 2014, the U.K. commissioned a study on the global problem of antimicrobial resistance. The study concluded that we must be concerned about drug resistance globally if we expect to be safe at home. An important theme discussed in the report is that infectious diseases profoundly affect poorer countries due to unsanitary living conditions. Unsanitary conditions, the report says, "act as a catalyst of rapid person-to-person spread, which can lead to an increase in the use and overuse of antibiotics."

Dr. Bell, what effect do poor living conditions and access to clean food and water have on the spread of infectious diseases, and how can poor sanitation contribute to the overuse of antibiotics?

Dr. BELL. Yes, thank you. This is unfortunately very much the case that—and you could imagine how, if you think about our prevention strategies in terms of infection control and in terms of hygiene how countries where people really just don't have the option of this sort of hygiene that we can—the level that we can establish here in the United States, but this does contribute to spread. And we see this in many countries around the world.

I think one of the things that we've learned is that I think it makes sense to sort of break some of this down in order to approach it. And infection control in hospitals, for example, is one component that we've really been working now with a lot of countries to at least improve infection control in facilities. If we see that, a lot of the most resistant bacteria are in facilities. We've been working with a number of countries in India, in West Africa— you heard with Ebola that's not about antibiotic resistance, but we certainly saw what the importance of infection control was.

Ms. CASTOR. So when poor countries address problems such as sanitation, access to clean water and infections in their medical facilities—this is basic—that helps reduce antibiotic resistance. And what more can be done then? We have so many fantastic nonprofit organizations, academics, churches, the developed world. What else can that group be doing to reduce the spread of infection?

Dr. BELL. Yes, you're absolutely right. And actually, I'll go back again to Ebola while that isn't a resistant organism. In our work in West Africa to improve infection control, we used many nongovernmental organizations, church groups, and other sorts of nongovernmental groups as trainers, as extenders, and as groups that can stay in the area and help the ministries of health and the hospitals with infection control.

Ms. CASTOR. So some experts have said and the U.K. report said and I think Mr. McKinley had said what needs to be done. They emphasized kind of some global organizing, collaborative or campaign. How could we go about such an effort?

Dr. BELL. The World Health Organization actually is in the process of organizing such a campaign, and while we all need to participate in that and provide some leadership.

Ms. CASTOR. So that would include tackling the over-the-counter prescriptions of antibiotics? The U.K. study reported that in parts of Southern and Eastern Europe, 20 to 30 percent of antibiotics are consumed without a prescription, and in parts of Africa, the figure rises to 100?

Dr. BELL. Yes. That's definitely part of what we have to tackle. Ms. CASTOR. OK. And then, Dr. Woodcock, Internet sales of antibiotics is a huge problem that this committee has documented extensively. You have testified before this committee multiple times on the use of Internet drugs. What role does FDA believe that Internet drug sales have on drug resistance across the globe?

Dr. WOODCOCK. I don't think we know in the United States. We intercept some packages and so forth. Most of the ones that we intercept are not antimicrobials. But certainly that could play a role. But I think that's something that's very difficult to get your hands around because, again, it's an international problem. Some of these are counterfeit, and that often plays a big role, especially in places like Africa or potentially even here because they don't have the right amount in there but they may have a low amount of antimicrobial in them, which then promotes the generation of resistance. So the free flow of antimicrobials, whether from over-the-counter use or through the Internet, is a potentially continuing large problem and very difficult to manage.

Ms. CASTOR. Thank you very much.

Mr. BURGESS [presiding]. The chair thanks the gentlelady. The gentlelady yields back. The chair recognizes the gentlelady from Tennessee 5 minutes for questions.

Mrs. BLACKBURN. Thank you, Mr. Chairman.

And I appreciate the patience you all have had with us this morning as we are running up and down to committees, hearings going on at the same time. So we have been in and out and you have been patient.

Dr. Hatchett, I want to come to you. Going back into February BARDA established the accelerator to support and research and development—and of course as we have worked on 21st Century Cures, we have been so interested in that, and then looking at the candidates that you were going to accelerate through this program with the drugs and vaccines and diagnostics and move that into development. And for us then moving it on to commercialization and into the mainstream.

I would like it if you could take just a minute and talk a little bit about the specific candidates that you all have put in that, about some of the vaccines or the microbials, some of the alternative therapies that you are reviewing in that program.

Dr. HATCHETT. Yes, ma'am. Thank you for the question. Just to be clear, the accelerator has not been established yet. We put out the solicitation earlier this year, and we have received the proposals. We had a good response, and we actually are in the process of negotiating with the lead candidates right now. So we haven't——

Mrs. BLACKBURN. OK.

Dr. HATCHETT [continuing]. Made the award——

Mrs. BLACKBURN. When you say a good response, quantify that for me a little bit.

Dr. HATCHETT. We had a pre-proposal conference where we had a—I don't remember the specific number, and I'm under oath, but I can get it for you.

Mrs. BLACKBURN. That would be great.

Dr. HATCHETT. But we ended up with five proposals, which really aggregated—actually some of the people that came to the pre-proposal conference met each other and then decided to partner, which made the proposals stronger. So I believe we had five proposals.

And one of the requirements of—I don't think it was a requirement, but we requested it. It was that people making a proposal to us could leverage the U.S. Government funding by gaining access to additional funding either from other funders or other venture capital entities, for example, that was going to be viewed as a positive. And fortunately, we had tremendous success with that, so we think when we make the final award, actually our investment, which this year could be as much as $30 million——

Mrs. BLACKBURN. OK.

Dr. HATCHETT [continuing]. May serve as a catalyst for additional investment coming in from other funders, even other countries that may be interested because of the way we've structured it. The—Ms. Castor mentioned the July 2014 U.K. study that was put together and led by Jim O'Neill, who was the former CEO of Goldman Sachs, and one of the recommendations of that study was to create a global innovation fund for antibiotic development. And actually, the United Kingdom and China have already made contributions of about $50 million to such a fund.

We think the accelerator will certainly bring in that level of funding. We are committed to providing up to $250 million for the accelerator over the next 5 years, and that could be a major, major catalyst for international collaboration to support this early-stage development.

Mrs. BLACKBURN. OK. And what is your expectation once you are able to populate the accelerator basically? Then how long do you think it will be before we begin to see next-generation products that are ready to go to the marketplace?

Dr. HATCHETT. So the accelerator, as I mentioned earlier, we are partnering very closely with our colleagues at NIAID. And NIAID has a full suite of what we call product development support services that can help innovators in the early stage accelerate their development. Our funding will allow those innovators to accelerate their timelines for bringing these products forward, and so we may be able to shorten the timelines. If those innovators were left to their own devices and left to the vagaries of the capital markets, it might take them many years to bring those products to the point that they would be ready for advanced development. The things that are in the accelerator are not going to pop out of the accelerator right into the marketplace. They're going to be brought to the stage of clinical development——

Mrs. BLACKBURN. Right. We——

Dr. HATCHETT [continuing]. Which could take several years.

Mrs. BLACKBURN. We understand that, but we——

Dr. HATCHETT. Right.

Mrs. BLACKBURN. When we look at some of these drugs that are taking 10 years, 12 years to get through the process, and you look at the entire compendium and you look at what goes on with them, the innovators working with the FDA, our hope is that we are going to see your process work and help to push this to the marketplace sooner, that there will be some efficiencies. We know you have got some challenges, but we are hopeful there will be some efficiencies in moving it forward.

Dr. HATCHETT. Ours too, ma'am. The perhaps relevant example—its slightly different scale of urgency—but when we have had to respond to events like the Ebola epidemic——

Mrs. BLACKBURN. Yes.

Dr. HATCHETT [continuing]. Or to the pandemic in 2009, we were able to push things forward with incredible velocity and so shorten normal development time frames. It might be 5, 7, 10 years down to even, you know, 9 months, a year, 2 years——

Mrs. BLACKBURN. OK.

Dr. HATCHETT. I'm not promising——

Mrs. BLACKBURN. That is helpful.

Dr. HATCHETT [continuing]. That we can do that with the accelerator——

Mrs. BLACKBURN. Right.

Dr. HATCHETT [continuing]. But that's the idea.

Mrs. BLACKBURN. That is helpful. That is what we want to hear.

Thank you. I yield back.

Mr. BURGESS. The gentlelady yields back. The chair thanks the gentlelady.

And the chair recognizes the gentleman from New York, Mr. Tonko, for 5 minutes, please.

Mr. TONKO. Thank you, Mr. Chair.

My good friend and colleague, Representative Louise Slaughter, to my knowledge the only microbiologist in Congress, has been raising alarm bells for quite some time about the excessive use of antibiotics in our farm animals and feedstock.

I know that the FDA has shared similar concerns about antibiotic use and resistance in farm animals. In fact, FDA's Center for Veterinary Medicine has developed a multipronged effort to limit or reverse resistance arising from the use of antibiotics in food-producing animals.

Dr. Woodcock, I understand that FDA's efforts to control antibiotic use in farm animals are not entirely in your wheelhouse given where you sit in the agency. However, can you walk us through what FDA is doing to address the nexus between antibiotic resistance and the overuse of antibiotics at the farm?

Dr. WOODCOCK. Yes. As I said in my oral testimony, the Center for Veterinary medicine has sought commitments from the manufacturers of animal drugs that have important medical uses, and they have all committed to submit supplements that would basically change these drugs to prescription only. And that means they would not be used for growth promotion purposes, and they would need to be prescribed by a veterinarian. The Center for Veterinary Medicine expects this to occur at the end of the year, the supplement submission. And so soon after that the changeover could be accomplished is my understanding.

Mr. TONKO. And with that in mind, in any way are we making progress in overuse of antibiotics in our food supplies?

Dr. WOODCOCK. That I can't answer but we could get back to you on that.

Mr. TONKO. Thank you. I would appreciate that. And, Dr. Woodcock, how are we tracking how much and what kinds of antibiotics food producers are using, and who is in charge of that effort? And just what is being done?

Dr. WOODCOCK. Well, in speaking to the Center for Veterinary medicine, I understand that tracking down to that level is difficult right now. We don't have a billing system similar to what we have for human drugs where we understand what prescriptions are issued for what animals and so forth. So my understanding is that that's a difficult set of information to find out.

Mr. TONKO. And I understand that one concern of regulators is that antibiotics have been used extensively for feed production purposes and not only for therapeutic purposes. So can you describe how antibiotics have been misused in that regard?

Dr. WOODCOCK. Well, there's been a long tradition as I understand—I'm not a veterinarian—but I understand there's been a long tradition of using certain amounts of antimicrobials preventively or in food that—in feed for the animals that results in some growth acceleration of the animals. And it's not treating an infection or anything like that. It's simply to put it in the feed and then the growth accelerates. And of course they're being produced as food animals, and so that is an economic benefit. That type of use is what's been addressed by the step that the Center for Veterinary Medicine is taking.

Mr. TONKO. And what are FDA and USDA planning to do to address that aspect of the problem?

Dr. WOODCOCK. Well, we expect the manufacturers will honor their commitments. They will submit supplements, and these will be changed.

They cannot be used for growth-promoting purposes, these medically important antibiotics.

Mr. TONKO. Are there any other efforts that can be made by FDA to control the use of antibiotics in our food supplies?

Dr. WOODCOCK. Well, I think there's always more that can be done, and I also believe that the norm system is very important and probably needs to continue to be strengthened. I think the value of surveillance just can't be overestimated because we need to know what's going on and at every level, from production all the way through to the food itself, that monitoring is extremely important for us to know what's going on.

Mr. TONKO. I take the efforts obviously are very important to the public——

Dr. WOODCOCK. Yes.

Mr. TONKO [continuing]. And we are at least pleased to hear that there is that kind of review being conducted, but I would love to see efforts go forward to make certain that there is every bit of safety and consumer-oriented response so that we can move forward progressively.

Dr. WOODCOCK. Thank you.

Mr. TONKO. With that, I yield back.

Mr. BURGESS. The gentleman yields back. The chair thanks the gentleman.

The chair recognizes the gentleman from Oklahoma, Mr. Mullin, 5 minutes for questions, please.

Mr. MULLIN. Well, I don't know about the antibiotics in cattle, but I live on a farm and I am on three different antibiotics as we speak. I literally just got off Bactrim and I am on a Z–Pak and I have got another one that was prescribed to me yesterday. And so to say that they are overprescribed, I thank the Lord for them, but I will probably say you are accurate on that.

With that being said, the University of Oklahoma has done a phenomenal job on trying to close the gap on what they call the superbug, and they recently developed a new antibiotic to go after what a lot of people refer to as MRSA or a staph infection. But even with all that development, Dr. Woodcock, despite the recent success at the University of Oklahoma, are there concerns that new drugs aren't being developed quickly enough to help close this gap?

Dr. WOODCOCK. Yes. There are major concerns because the microbes are kind of like the criminals. They're always one step ahead of us, right, and if you expose people and you expose the bacteria to an antimicrobial, some of them will develop resistance. So we need a robust pipeline.

We have approved a number of antibacterials recently for MRSA infections, new ones, but the general pipeline of antimicrobials is not robust. It's very fragile. We don't just need very targeted antimicrobials, we need a broad set of antimicrobials for the future. And that's still not happening, although there—you know, certainly there are discoveries being made.

Mr. MULLIN. So with the current crop of antibiotics out there, do you see where we are going to be able to catch up or we are going to continue to stay behind or get farther behind?

Dr. WOODCOCK. I think we need to be very concerned because it's not the work of a year to catch up, it's the work of decades. And these development programs—Dr. Burgess mentioned Pfizer earlier; they're no longer in the space of development. These development programs were terminated by lots of companies several decades ago.

Mr. MULLIN. Due to?

Dr. WOODCOCK. Lack of economic incentives and the fact that at that time there was of course a broad range of antimicrobials available. And even though we could foresee resistance occurring in the future, the return on investment wasn't there compared to, say, cancer or other fields.

Mr. MULLIN. Well, Dr. Woodcock, how long does it take, say, a company that is wanting to develop this, how long does it take them from when they start to when they can probably bring a product to the market?

Dr. WOODCOCK. If you start from discovery, probably 10 years.

Mr. MULLIN. At the cost of what?

Dr. WOODCOCK. The estimates are very controversial. Many people have said between $1 billion and $2 billion for a new molecular entity.

Mr. MULLIN. So is there a way to speed this up? I mean, if we are behind because, my goodness, I can understand why a company would be hesitant to invest $1–2 million on an——

Dr. WOODCOCK. Billion.

Mr. MULLIN. Billion, I am sorry, $1–2 billion on an antibiotic and it takes 10 years to bring it to market and the success rate is, I am sure, pretty low. Do you know what the success rate is on that?

Dr. WOODCOCK. For antimicrobials, it's—if you start from discovery, it's going to be one in many thousands, OK——

Mr. MULLIN. One in many thousands, so——

Dr. WOODCOCK. You start into entry into the clinic, it might be 1 in 10.

Mr. MULLIN. So what is the incentive for a company to do this?

I mean, I am thinking as a business owner I am going, OK, I am going to invest $1–2 billion and I am going to have 1 out of 1,000— let's just use that number—of a chance of it actually coming to market. Is there an incentive? Is there a way to bring this number down? Is there a way to help improve that?

Dr. WOODCOCK. Well, I would also add that we would urge you as a business owner once you develop this drug to make sure it's used very narrowly and not widely used because otherwise resistance would develop.

Mr. MULLIN. Well, I agree with that, but you have got to get your money back.

Dr. WOODCOCK. I understand that.

Mr. MULLIN. I mean, $1–2 billion——

Dr. WOODCOCK. I'm just saying that's the market—that's the return-on-investment problem. What Dr. Hatchett and Dr. Dixon have been talking about are Federal Government efforts to accelerate this pipeline and ease the barriers to getting so it doesn't cost so much and the success rate could be higher.

Mr. MULLIN. Is the FDA looking into this to help ease those barriers, help speed that process up? Because we can all tell this is going to be a major problem, and so it seems like we should be working together on this.

Dr. WOODCOCK. Oh, yes. And we've been doing things for years.

We have issued a number of new guidances on new pathways.

We're working with outside public-private partnerships on new end points, different ways to study the drugs, and then what was mentioned earlier, the limited population use that's in the 21st Century Cures would be another way of streamlining the development programs.

Mr. MULLIN. Dr. Woodcock, thank you. I am out of time.

Mr. BURGESS. The gentleman yields back. The chair thanks the gentleman.

The chair recognizes the gentlelady from Illinois, Ms. Schakowsky, 5 minutes for questions.

Ms. SCHAKOWSKY. This is exactly the conversation that I have been wanting to get into.

Dr. Woodcock, in conclusion you say "It is virtually undisputed that we are facing a public health crisis because of the rise of serious resistant infections and the simultaneous decline in R&D in this area." What I am hearing is that there is not enough profit in addressing this problem. And that sets my hair on fire. If this is a worldwide public health issue and the reason that we cannot make progress is because—as have been called—our industry partners are unwilling to do that because there is not enough money, then it seems to me exactly the space then that government needs to step in and deal with this.

I just don't understand the word "incentives"—and there may be other ways besides money. I don't know. But what are we talking about here, that these companies are getting out of the business, that the Pfizers are getting out of the business. There is just not enough dough here when at some point this could be just a worldwide problem of the bug that is going to kill, I don't know, millions of people. Somebody answer me.

Dr. BELL. Well, I would say there are two problems. One is it's very hard to discover and develop these——

Ms. SCHAKOWSKY. Right.

Dr. BELL [continuing]. OK, scientifically. And then once you get them on the market, we're going to—everyone will ask them is don't sell them, OK, don't sell them very much because they're too valuable to waste on—that was the paradigm in the past, and that's why antimicrobials are so overprescribed, OK, is they were used for minor infections——

Ms. SCHAKOWSKY. Well, can I——

Dr. BELL [continuing]. And viruses——

Ms. SCHAKOWSKY [continuing]. Interrupt for a second? I think one of the other issues that contributes to that problem is TV advertising of drugs, which has created an atmosphere among consumers that I am in charge of my health, I am going to go to my doctor, I am going to say what I want, I will find a way to get that drug. I just wanted to add that. I think that is part of the culture that contributes to this problem. Anybody?

Dr. HATCHETT. Thank you. It's a critically important question. And I have seen an estimate that the market for antibiotics in the United States is about $40 billion. But of that $40 billion for that market, only about $4–5 billion is for drugs that are unpatented. So any new drug that enters the market enters a market where there are dozens of competing generic antibiotics. And so the new entries can't charge a premium unless they can fully differentiate themselves from all of the other antibiotics that are on the market. And so it suppresses the ability to achieve profit for a new drug.

And you've heard the estimates of what it costs to bring a new drug to market, and so companies, to recoup their investment over many, many years have to see opportunities in the marketplace that they can achieve that return even almost just to a breakeven.

The model that BARDA has implemented in other areas where the market is failing to deliver public health requirements so against agents of bioterrorism or pandemic influenza, for example, we over the past decade have evolved a model where we provide substantial advanced research and development funding for products that have reached the stage of clinical development. So these are things that are being tested in human clinical trials. We also provide a market entry incentive, a pull incentive in terms of a procurement of the product. So for something that we're developing for bioterrorism, we then buy a large quantity of it and put it into the strategic national stockpile so a person working under it has a guaranteed market commitment.

And the other thing that we provide, which is very, very critical, we found this to be extremely critical is we provide access to a core of experts on all aspects of product development who can assist potentially smaller companies that don't have all of this expertise inhouse and to access to product development infrastructure. And so it's a multi-legged stood in terms of the support that we provide.

It wasn't until we put all of those components together for our bioterrorism threats, for pandemic influenza threats that we really started to see our program succeed. We think many of those elements are going to be required to overcome these adverse market courses that are leading to disinvestments.

Ms. SCHAKOWSKY. Well, I hope that bottom line isn't that we have to look at the bottom line of for-profit companies to figure out whether we are going to protect the health of this planet.

Mr. BURGESS. The gentlelady yields back.

Let me if I could, I would like to ask a follow-up question because the Center for Medicare and Medicaid services yesterday put out a relatively lengthy proposed rule on the stewardship program. You have had a chance to familiarize yourself with the proposed rule that CMS put out. Now, the question is, once again, do we get the right balance between the regulatory burden and effective governance? So, again, I would just appreciate your thoughts on the proposed rule.

Dr. BELL. I haven't actually seen the rule itself, Congressman, but we have been working with CMS around this concept for quite some time. They have already actually had previously issued some comments for similar kind of conditions of participation for longterm care. Now, this is the one for acute care.

So I appreciate your point about a balance between regulation and individual clinical judgment. In general, these conditions of participation, they've been working closely with us and they basically are meant to incorporate the core principles of stewardship that we published in 2014 for inpatient and 2015 for long-term care. So these are really just fundamental kind of necessary pieces of a puzzle so that a facility has a stewardship program that actually is able to influence the process.

And I will say also that this whole process of coming up with the core elements and CMS's participation has really been very broadly supported in general by American Hospital Association and by lots of kind of industry partners. I think there really is a broad agreement now that the concept of helping physicians and facilities and patients prescribe wisely is really pivotal.

We've been talking a lot about drug development and what all the gaps are there, but really—and, as you've heard, I think this whole issue of antibiotic resistance is a long-term problem that has many components, and we really have to address them all in order to tackle it. And one of the pivotal components is this issue of stewardship, of getting serious about doing something about overprescribing, not stopping people from prescribing but being able to preserve the antibiotics that we have now. And that's sort of about saving lives now, as well as these areas of prevention and diagnostics and new drug development.

So as I say, I'm not familiar with the specifics, but it has been really a very broad process that CMS has gone through and has collaborated quite closely with us.

Mr. BURGESS. Dr. Woodcock, on the issue of stewardship—and you have mentioned in your opening statement in response to Mr. Tonko's question about involving the USDA and the veterinarian space in this. It is one thing to come down hard on the private practice doc in the United States for overprescribing antibiotics, and I probably include myself in that group that would feel put upon by some of these directions. But honestly, we can't compete with what is being used in feedlots.

Dr. WOODCOCK. Yes. Well, as I said, the Center for Veterinary Medicine has taken steps around that and secured commitments from those manufacturers of medically important antimicrobials that they will submit a supplement that will eliminate that use basically.

Mr. BURGESS. And let me just——

Dr. BELL. If I could just add that the concept of stewardship and stewardship for veterinarians, in addition to stewardship for human doctors is something that the American Veterinary Medical Association has been quite interested in, and we've been providing them actually with a lot of tools that they could use with veterinarians. And actually we haven't spoken about companion animals, but this is another area actually where we really don't have a very good sense——

Mr. BURGESS. Sure.

Dr. BELL [continuing]. Of what's happening, and we will be——

Mr. BURGESS. Let me——

Dr. BELL [continuing]. Looking to establish something there.

Mr. BURGESS. I am going to stop there for—I want to ask one other question, Dr. Woodcock, since we are speaking about animal products. The issuance under an emergency use authorization for a genetically modified mosquito in parts of the world that are affected by the Zika virus, why is that taking so long? Why is it so difficult? Why is a genetically modified mosquito having to go through a new drug application?

Dr. WOODCOCK. Well, that is out of my wheelhouse, and we'd have to get back to you on that particular question.

Mr. BURGESS. OK.

Dr. WOODCOCK. It's not a drug——

Mr. BURGESS. But my understanding, it is being treated as if it was a new drug application, so I appreciate that is not part of today's discussion, but I really do want to follow up with you on that, because I think it is an important issue.

Seeing no further members wishing to ask—you have a question? The gentlelady from Illinois just wants me to conclude, so I do want to thank all our witnesses and members who have participated in today's hearing. I remind members they have 10 business days to submit questions for the record, and I ask all the witnesses to agree to respond promptly to these questions.

With that, the subcommittee stands adjourned.

[Whereupon, at 12:01 p.m., the subcommittee was adjourned.]

[Material submitted for inclusion in the record follows:]

U.S. House of Representatives, Committee on Energy and Commerce, June 10, 2016

TO: Members, Subcommittee on Oversight and Investigations
FROM: Committee Majority Staff
RE: Hearing entitled "Combatting Superbugs: U.S. Public Health Responses to Antibiotic Resistance"

The Subcommittee on Oversight and Investigations will hold a hearing on Tuesday, June 14, 2016, at 10:00 a.m. in 2322 Rayburn House Office Building, entitled "Combatting Superbugs: U.S. Public Health Responses to Antibiotic Resistance." The Subcommittee will hear testimony from U.S. public health officials on the current risks associated with "superbugs" resistant to antibiotics and the Federal government's plans to confront this public health challenge.

Witnesses

- Dr. Beth Bell, Director, National Center for Emerging and Zoonotic Infectious Disease, Centers for Disease Control;
- Dr. Janet Woodcock, Director of the Center for Drug Evaluation and Research, Food and Drug Administration;
- Dr. Richard Hatchett, Acting Director, Biomedical Advanced Research and Development Authority; and
- Dr. Dennis Dixon, Division of Microbiology and Infectious Diseases, National Institute of Allergy and Infectious Diseases, National Institutes of Health.

Background

Understanding the Superbug and Antibiotic Resistance

One of the world's most pressing health problems is the emergence of bacterial infections that are resistant to antibiotics.[4] Since the discovery of penicillin in the early 20th century, almost every type of bacteria is becoming resistant to the antibiotic treatment designed to treat it. The continuing evolution of bacteria, the over-prescription of antibiotics, and the lack of new antibiotic development have contributed to this problem.

According to the Centers for Disease Control and Prevention (CDC), at least 2 million Americans fall sick every year with antibiotic-resistant infections, and 23,000 die.[5] Globally, some institutions estimate up to 700,000 die each year from antibiotic resistant infections.[6] CDC Director Thomas Frieden has commented that "[t]he medicine cabinet is empty for some patients . . . . It is the end of the road unless we act urgently."[7] Without action, the researchers estimate 10 million people will die per year by 2050 from drug resistant infections.[8]

Medical professionals commonly define a "superbug" as a bacterial infection that is resistant to most or all antibiotics. Outbreaks of antibiotic-resistant bacteria occur most often in hospital settings, among patients receiving care for serious conditions. One of the most recent and high profile outbreaks occurred at the National Institutes of Health (NIH) Clinical Center. In 2012, the NIH announced that a type of Carbapenem-Resistant Enterobacteriaceae (CRE) had struck its clinical center, infecting 19 patients and killing seven.[9] Over 70 bacteria are classified as CRE,

---

[4] The World Health Organization, *Antibiotic Resistance Fact Sheet*, October 2015, *available at* http://www.who.int/mediacentre/factsheets/antibiotic-resistance/en/.

[5] Centers for Disease Control and Prevention, *Antibiotic/Antimicrobial Resistance, available at:* https://www.cdc.gov/drugresistance/ (last visited June 8, 2016).

[6] Review on Antimicrobial Resistance, A*ntimicrobial Resistance: Tackling a Crisis for the Health and Wealth of Nations*, December 2014.

[7] Jen Christensen and Debra Goldschmidt, *A Dreaded Superbug Found for the First Time in a U.S. Woman*, CNN, May 27, 2016.

[8] Review on Antimicrobial Resistance, A*ntimicrobial Resistance: Tackling a Crisis for the Health and Wealth of Nations*, December 2014.

[9] Brian Vastag and Lena H. Sun, *NIH Superbug Claims 7th Victim*, THE WASHINGTON POST, Sept. 14, 2012.

including E.coli, and they typically exist in the digestive system. Over time, these bacteria have become resistant to last-resort antibiotics.

Dr. Frieden has described CRE as a "triple threat" because (1) the bacteria are resistant to all or nearly all antibiotics, (2) the bacteria kills up to half of patients who get bloodstream infections and (3) the bacteria can transfer their antibiotic resistance to other bacteria within the family, potentially making other bacteria untreatable.[10] For example, according to Dr. Frieden, CRE can "spread the genes that destroy our last antibiotics to other bacteria, such as E. coli, and make E. coli resistant to antibiotics also."[11] This is significant because E. coli is the most common cause of urinary tract infections in healthy people.[12] According to data reported to the CDC, the percentage of CRE resistant to antibiotics increased from 1.2 percent in 2001 to 4.2 percent in 2011.[13]

The issue of antibiotic resistance is again in the news because of a recent discovery of new antibiotic-resistant gene in Pennsylvania. Last month, a woman in Pennsylvania was diagnosed with an E. coli infection that had a rare gene called MCR-1, which makes it resistant to colistin, an antibiotic of last resort.[14] This antibiotic is used to treat patients with multiple drug-resistant infections, including CRE. However, colistin has very strong side effects and can cause kidney damage, so it is used in human patients only when absolutely necessary.[15]

The MCR-1 gene exists on a "plasmid," which is a small piece of DNA that can move from one bacteria to another, spreading antibiotic resistance.[16] In November 2015, Chinese and British researchers discovered the colistin-resistant strain in pigs, raw pork, and in a small number of people in China. The case of the woman Pennsylvania is the

---

[10] Lena H. Sun, *CDC Says 'Nightmare Bacteria' a Growing Threat*, THE WASHINGTON POST, March 5, 2013.

[11] *Id.*

[12] *Id.*

[13] Centers for Disease Control and Prevention, *Morbidity and Mortality Weekly Report, Vital Signs: Carbapenem-Resistant Enterobacteriaceae*, March 8, 2013.

[14] Lena H. Sun and Brady Dennis, *The Superbug That Doctors Have Been Dreading Just Reached the U.S.*, THE WASHINGTON POST, May 27, 2016.

[15] Colistin is not used at all in farm animals in the U.S.

[16] The Centers for Disease Control and Prevention, *Discovery of First MCR-1 gene in E. coli Bacteria Found in a Human in United States, Media Statement*, May 31, 2016.

first discovery of the rare MCR-1gene in the United States. How the woman contracted the rare gene is unknown, but the CDC is currently investigating in coordination with the Department of Defense (DOD). The woman did not have CRE, and was eventually treated with other antibiotics and was released from medical supervision.

This discovery is troubling to health officials because it signals the potential arrival of an unstoppable superbug. The MCR-1 gene can join with a more common superbug, such as CRE, to create a bacterial infection that cannot be stopped with any known antibiotic treatment, which is known as a pan-resistant infection. In response to the discovery of the MCR-1 gene in the Pennsylvania case, Dr. Frieden commented that "[i]t basically shows us that the end of the road isn't very far away for antibiotics—that we may be in a situation where we have patients in our intensive care units, or patients getting urinary-tract infections for which we do not have antibiotics."[17] At present, however, there is no evidence that the MCR-1 gene has merged with CRE to form a pan-resistant infection.

Overuse of Antibiotics

The over-use of antibiotics when they are not completely necessary is a major contributor to the growing problem of antibiotic-resistance. As soon as an antibiotic goes in to wide use among the general public, bacteria will evolve to become resistant. This happened with penicillin in the 1940s, when it became commonly prescribed to treat the general public.[18]

A study published last month in the Journal of the American Medical Association (JAMA) quantifies the overuse of antibiotics in the United States. The study found that nearly a third of antibiotics prescribed in doctors' offices, emergency rooms, and hospital-based clinics in the United

---

[17] Lena H. Sun and Brady Dennis, *The Superbug That Doctors Have Been Dreading Just Reached the U.S.*, THE WASHINGTON POST, May 27, 2016.

[18] Julian Davies and Dorothy Davis, *Origins and Evolution of Antibiotic Resistance*, Microbiol Mol Biol Rev, Sept. 2010.

States are not needed.[19] This amounts to nearly 47 million unnecessary prescriptions given out each year.[20] Most of the unnecessary antibiotics are prescribed to treat respiratory conditions such as the common cold, bronchitis, and other viral illnesses. Antibiotics are not effective against viral illnesses, but doctors sometime prescribe antibiotics anyway when they cannot determine whether the infection is bacterial or viral.

The CDC and the Pew Charitable Trust collaborated on this study, which analyzed CDC data for all antibiotic use in 2010 and 2011 in doctors' offices, emergency rooms, and hospitalbased clinics.[21] The numbers in the report most likely undercount the use of antibiotics, because the data did not include urgent care clinics, retail pharmacies, dentists' offices, and prescriptions given over the phone, and by nurse practitioners and physician assistants.[22] About 13 percent of all outpatient visits in the United States result in an antibiotic prescription, which amounts to about 154 million visits annually.[23]

Dr. Frieden has commented that over-use of antibiotics will lead to undesirable consequences:

> Antibiotics are life-saving drugs, and if we continue down the road of inappropriate use we'll lose the most powerful tool we have to fight life-threatening infections . . . . Losing these antibiotics would undermine our ability to treat patients with deadly infections [and] cancer, provide organ transplants and save victims of burns and trauma.[24]

---

[19] *Prevalence of Inappropriate Antibiotic Prescriptions Among US Ambulatory Care Visits, 2010-2011*, THE JOURNAL OF THE AMERICAN MEDICAL ASSOCIATION, May 3, 2016.

[20] *Id.*

[21] *Prevalence of Inappropriate Antibiotic Prescriptions Among US Ambulatory Care Visits, 2010-2011*, THE JOURNAL OF THE AMERICAN MEDICAL ASSOCIATION, May 3, 2016.

[22] Pranita D. Tamma and Sara E. Cosgrove, *Addressing the Appropriateness of Outpatient Antibiotic Prescribing in the United States: An Important First Step*, THE JOURNAL OF THE AMERICAN MEDICAL ASSOCIATION, May 3, 2016.

[23] *Prevalence of Inappropriate Antibiotic Prescriptions Among US Ambulatory Care Visits, 2010-2011*, THE JOURNAL OF THE AMERICAN MEDICAL ASSOCIATION, May 3, 2016.

[24] The Centers for Disease Control and Prevention, *CDC: 1 in 3 Antibiotic Prescriptions Unnecessary, Press Release*, May 3, 2016.

Although antibiotic overuse is commonplace in the United States, it is even more egregious across the globe. The World Health Organization has declared that humanity is on the precipice of a "post-antibiotic era," where common infections may once again be lethal because bacteria have become resistant to the antibiotics existing to treat them.[25] The use of colistin, the antibiotic of last resort, is more common in Europe and Asia, and several recent studies from multiple countries show the emergence of colistin-resistant bacteria throughout the world.[26]

Development of New Antibiotics

Although reducing the inappropriate and unnecessary use of antibiotics will slow the ability of bacteria to become resistant to known antibiotics, it alone will not solve the problem. New antibiotics must be developed. However, there are well-documented barriers to the discovery and development of new antibiotics.

The last 30 years have marked a significant reduction in the discovery on new antibiotics. Currently, every antibiotic is a derivative of a type of antibiotic discovered from 1900 to 1984.[27] After the rush of discoveries of new antibiotics in the 1950s, discovery of new compounds as a foundation for antibiotics became more challenging for scientists, and new antibiotics became rarer over time. FDA approvals for antibiotics fell from 29 during the 1980s to just nine from 2000 to 2010.[28]

In addition to the poor discovery prospects and gaps in scientific research, antibiotics are expensive to develop and offer a poor return on investment.[29] Antibiotic prescriptions are supposed to be limited to reduce bacterial resistance, and a new antibiotic on the market would be guarded closely. Doctors may be reticent to prescribe a new antibiotic that fights antibiotic-resistant infections, except in rare circumstances, because bacteria will become resistant to that antibiotic as soon as it becomes

---

[25] World Health Organization, *Antimicrobial Resistance: Global Report on Surveillance*, 2014.

[26] The Pew Charitable Trusts, *A Scientific Roadmap for Antibiotic Discovery*, May 2016.

[27] *Id.*

[28] The Pew Charitable Trusts, Infectious Diseases Society of America, Pharmaceutical Research and Manufacturers of America, *Reviving the Pipeline of Life-Saving Antibiotics: Exploring Solutions to Spur Innovation*, Conference proceedings, Sept. 22, 2011.

[29] The Pew Charitable Trusts, *A Scientific Roadmap for Antibiotic Discovery*, May 2016.

commonly used. Moreover, antibiotics are generally low-cost drugs designed to treat acute illnesses, so it is difficult for companies to make a profit.

An analysis by the Pew Charitable Trusts tracks new antibiotics in development, and found the current crop is insufficient to meet current or anticipated patient needs. In addition, few of the antibiotics in development can address the most antibiotic-resistant infections. As of March 2016, there were 37 new antibiotics in development. Of the 37 antibiotics in development, 11 were in phase 1 clinical trials, 13 in phase 2, and 13 in phase 3.[30] Historically, about 60 percent of drugs that enter phase 3 will be approved.[31] These drugs would potentially address many, but not all, resistant bacteria.

Of the approximately 34 companies with antibiotics in clinical development, only five are in the top 50 pharmaceutical companies by sales data.[32] About half of those companies are small companies that have no products on the market.[33]

The Federal Response to Antibiotic Resistance

Public health officials have long warned about the risks of antibiotic resistance, and Congress and executive agencies have responded to these risks. Led by the Department of Health and Human Services (HHS), numerous Federal agencies have joined the effort to reduce the threat of antibiotic resistant bacteria. The CDC, the Food and Drug Administration (FDA), the NIH, and the Biomedical Advanced Research and Development Authority (BARDA) have all made significant and ongoing contributions to thwart the spread of antibiotic resistant bacterial infections.

---

[30] The Pew Charitable Trusts, Tracking the Pipeline of Antibiotics in Development, March 2014, updated May 2016, *available at* http://www.pewtrusts.org/en/research-and-analysis/issue-briefs/2014/03/12/tracking-the-pipeline-of-antibiotics-in-development.

[31] Michael Hay et al., *Clinical Development Success Rates for Investigational Drugs*, Nature Biotechnology 32 (2014).

[32] The Pew Charitable Trusts, Tracking the Pipeline of Antibiotics in Development, March 2014, updated May 2016, available at http://www.pewtrusts.org/en/research-and-analysis/issue-briefs/2014/03/12/tracking-the-pipeline-of-antibiotics-in-development.

[33] *Id.*

Congress has acted to encourage the innovation and discovery of new antibiotics. As part of the Food and Drug Administration Safety and Innovation Act, the Generating Antibiotic Incentives Now (GAIN) Act was signed into law in 2012. The purpose of the GAIN Act was to promote the development and expedite the FDA approval process of antibiotics to treat lifethreatening infections. To promote antibiotic development, the GAIN Act added an additional 5 years of exclusivity, which means that a generic form of the antibiotics may not be produced for an additional 5 years. This allows drug companies extra time to recoup productions costs and increases the incentive for drug companies to research and develop new antibiotics. The Antibiotic Development to Advance Patient Treatment (ADAPT) Act builds on this progress, and passed through the House as part of the 21st Century Cures legislation last year. The ADAPT legislation aims to encourage development of antibiotics for life threatening bacterial infections and provides an alternative regulatory pathway for limited-population antibiotics.

In March 2015, the White House released a National Action Plan for Combating Antibiotic-Resistant Bacteria (CARB) in response to an Executive Order issued by President Obama in September 2014.[34] This action plan outlines steps to implement a national strategy to combat antibiotic resistance and addresses policy recommendations made by the President's Council of Advisors on Science and Technology.

The National Action Plan listed five goals:

- Slow the emergence of resistance bacteria and prevent the spread of resistant infections;
- Strengthen national one-health surveillance efforts to combat resistance;
- Advance development and use of rapid and innovative diagnostic tests for identification and characterization of resistant bacteria;
- Accelerate basic and applied research and development for new antibiotics, other therapeutics and vaccines; and

---

[34] The White House, *National Action Plan for Combating Antibiotic-Resistant Bacteria*, March 2015.

- Improve international collaboration and capacities for antibiotic-resistance prevention, surveillance, control and antibiotic research and development.

The plan also created a Task Force comprised of representatives from the Federal agencies charged with implementing the plan.[35] Congress has increased funding for these initiatives by 57 percent over last fiscal year, for a total of more than $375 million.[36]

The action plan states that by 2020, implementation of the plan will lead to major reductions in the "incidence of urgent and serious threats" from antibiotic-resistant bacteria, "improved antibiotic stewardship in healthcare settings," "expanded surveillance for drug-resistant bacteria in humans and animals," and the "development of two or more antibiotic drug candidates or non-traditional therapeutics for treatment of human disease."[37]

In November 2015, the Task Force released its "First 180 Days Report" monitoring the progress of the action plan. The report highlighted notable achievements, including new antibiotic stewardship guidelines for nursing homes from the CDC.[38] In March 2016, the President's Advisory Council on Combating Antibiotic-Resistant Bacteria issued "initial assessments" of the progress on the action plan.[39]

Two of the Council's six recommendations addressed the need for Federal coordination across agencies. The Council noted that "centrally coordinated mechanisms were not sufficient to ensure maximum synergy, avoidance of duplication, and coverage of all key points."[40] The Council also recommended selecting a "champion in the [U.S. Government] to

---

[35] *Id.*

[36] For the 2016 fiscal year, Congress allocated $160 million to the Centers for Disease Control and Prevention, $100 million to the National Institutes of Health, and $96 million to BARDA.

[37] The White House, *National Action Plan for Combating Antibiotic-Resistant Bacteria*, March 2015.

[38] Taskforce for Combating Antibiotic Resistant Bacteria, *First 180 Days Report*, November 2015.

[39] Presidential Advisory Council on Combating Antibiotic-Resistant Bacteria, *Initial Assessments of the National Action Plan for Combating Antibiotic-Resistant Bacteria*, March 2016.

[40] *Id.*

align all of the agencies and move the work forward efficiently and synergistically."[41] These recommendations were made to the HHS Secretary, to advise and support the implementation of the CARB Action Plan.

Mapping the Spread of Antibiotic-Resistant Bacteria

The CDC has led the coordinated effort between the DOD and the U.S. Department of Agriculture (USDA) to respond to the most recent discovery of MCR-1 in Pennsylvania. The CDC is currently identifying and investigating the Pennsylvania woman's close contacts to try to identify whether any of them was the source of the MCR-1 gene. According to a briefing by the CDC provided to committee staff, unconfirmed testing of samples from the 20 high-risk contacts so far do not show the MCR-1 gene. The USDA has also been investigating the source of the MCR-1 gene, and discovered the gene in a pig in the United States. The USDA is currently working to identify the source of that gene as well.[42]

These efforts are made possible by the National Antimicrobial Resistance Monitoring System program, which monitors antimicrobial resistance across bacteria discovered in food, animals, humans, and meats. The FDA, CDC, and USDA all conduct research on bacteria found through these surveillance methods, to determine how the resistance arises and transfers between bacteria.[43]

As part of the CARB Action Plan, CDC is ramping up its network of regional and local labs to track the spread of the MCR-1 gene and other new forms of antibiotic resistance. Set to begin in fall of 2016, the "Antibiotic Resistance Lab Network" will provide the capacity for seven to

---

[41] *Id.*

[42] The Department of Health and Human Services, *Blog Post: Proactive Efforts by U.S. Federal Agencies Enable Early Detection of New Antibiotic Resistance*, May 26, 2016.

[43] The Food and Drug Administration, National Antimicrobial Resistance Monitoring System, *available at* http://www.fda.gov/AnimalVeterinary/SafetyHealth/AntimicrobialResistance/NationalAntimicrobialResistanceMon itoringSystem/ (last visited June 8, 2016).

eight regional labs charged with detecting and responding to bacteria-resistant organisms, and reporting these instances to the CDC.[44]

Antibiotic Stewardship Programs

In addition to assembling new data to track the use of antibiotics and the spread of these resistant bacteria, CDC has partnered with FDA to advocate for antibiotic "stewardship" programs in health care facilities throughout the United States.[45] The CDC has issued guidelines about how hospitals can minimize inappropriate or excessive use of antibiotics, which could reduce antibiotic over-prescription.[46] Not all health care facilities have implemented stewardship programs, but the Centers for Medicare and Medicaid Services is expected to release a proposed rule that would require hospitals to have antibiotic stewardship programs in place before they can receive reimbursements from Medicare and Medicaid.[47]

New Antibiotic Development

Scientists and medical professionals have noted that the current crop of candidate antibiotics in development is insufficient to counter the current threat of antibiotic resistant bacteria. In May 2015, there were 28 antibiotics in Phase II/Phase III clinical development, compared to over 500 candidates in Phase II/Phase III clinical development for oncology indications.[48] BARDA is responsible for developing and procuring medical countermeasures to address public health threats, including bacterial infections caused by antibiotic resistant bacteria.

---

[44] Centers for Disease Control and Prevention, *Antibiotic Resistance Solutions Initiative*, *available at* http://www.cdc.gov/drugresistance/solutions-initiative/ (last visited June 8, 2016).

[45] U.S. Food and Drug Administration, *Get Smart: Know When Antibiotics Work Campaign*, *available at* http://www.cdc.gov/getsmart/community/materials-references/index.html (last visited June 8, 2016).

[46] The Centers for Disease Control and Prevention, *Core Elements of Hospital Antibiotic Stewardship Programs*, 2014.

[47] Presidential Advisory Council on Combating Antibiotic-Resistant Bacteria, *Initial Assessments of the National Action Plan for Combating Antibiotic-Resistant Bacteria*, March 2016.

[48] Office of the Assistant Secretary for Preparedness and Response, *BARDA Seeks to Launch a Novel Partnership, A Product Accelerator to Address Antimicrobial Resistance*, Feb. 19, 2016, *available at* http://www.phe.gov/ASPRBlog/pages/BlogArticlePage.aspx?PostID=176.

In February 2016, BARDA collaborated with the National Institutes of Allergy and Infectious Diseases (NIAID) to establish a "Biopharmaceutical Accelerator" that will "support research and development to accelerate candidate products (drugs, vaccines, and diagnostics) into clinical development."[49] This program furthers the goals set out in the Administration's CARB Action Plan to incentivize antibacterial drug development. According to BARDA, the Accelerator will (1) fund development of antibacterial products, (2) quickly move successful drug candidates through early development, (3) provide business and drug development guidance, and (4) decrease barriers to research and development of antibiotics.[50]

The NIAID has been funding and conducting research on antimicrobial resistance, including basic research on how bacteria develop resistance, diagnostics, and clinical trials to find vaccines and treatments that are effective against antibiotic-resistant bacteria.[51] NIH has also funded studies to evaluate alternative therapies to traditional antibiotics.

Issues

The following issues will be examined at the hearing:

- The status of the current threats of antibiotic-resistant bacteria in the United States and around the world;
- The coordination among agencies on activities to combat antibiotic-resistant bacteria;
- The development of new antibiotics and alternative therapies to treat bacteria that are resistant to all or nearly all antibiotics; and
- The role of the Congress, and the Energy and Commerce Committee in particular, in shaping the response to antibiotic-resistant bacteria.

---

[49] *Id.*

[50] *Id.*

[51] National Institute of Allergy and Infectious Diseases, *Antimicrobial Resistance*, March 14, 2016, available at https://www.niaid.nih.gov/topics/antimicrobialresistance/Pages/default.aspx.

INDEX

A

access, 26, 32, 33, 67, 75, 86, 89, 90, 91, 95, 98, 99, 102, 105, 109, 110, 134, 135, 138, 141, 142, 143, 145, 204, 206, 233, 234, 236, 245
agencies, vii, 3, 5, 6, 11, 20, 21, 24, 32, 33, 40, 46, 47, 48, 50, 51, 53, 68, 76, 77, 83, 87, 93, 94, 95, 98, 100, 102, 105, 111, 112, 113, 114, 116, 117, 119, 123, 126, 127, 134, 135, 137, 139, 157, 160, 162, 163, 169, 171, 172, 185, 188, 191, 203, 222, 254, 256, 259
agency actions, 11, 22, 122, 127
alternative therapies, 158, 161, 169, 224, 235, 259
antibiotic resistance, v, vii, 1, 3, 4, 6, 7, 8, 9, 10, 11, 12, 14, 15, 16, 20, 21, 22, 23, 24, 25, 26, 27, 28, 29, 30, 31, 34, 40, 41, 42, 44, 45, 46, 48, 49, 50, 51, 52, 53, 54, 57, 59, 61, 66, 67, 68, 74, 83, 87, 94, 96, 99, 100, 101, 103, 106, 110, 111, 112, 114, 115, 117, 119, 120, 122, 123, 124, 125, 126, 127, 130, 134, 135, 136, 139, 141, 142, 143, 145, 155, 156, 157, 158, 160, 161, 162, 163, 164, 165, 166, 167, 168, 170, 171, 175, 176, 177, 178, 179, 180, 181, 182, 183, 184, 185, 187, 189, 193, 198, 202, 204, 205, 206, 207, 208, 209, 210, 211, 212, 213, 219, 220, 222, 223, 227, 230, 234, 238, 246, 248, 249, 250, 251, 254, 255, 257, 258
antibiotic-resistant infections, vii, 4, 6, 9, 10, 15, 16, 21, 22, 23, 24, 31, 42, 44, 45, 48, 51, 70, 71, 72, 76, 77, 78, 82, 87, 93, 94, 115, 118, 120, 122, 127, 134, 136, 157, 159, 176, 202, 206, 210, 212, 249, 253, 254
antibiotic-resistant superbugs, 156
antimicrobial therapy, 128
antiviral drugs, 189, 200
appropriate antibiotic use, vii, viii, 3, 7, 98, 101, 103, 105, 114, 137, 138, 140, 141, 143

B

bacteria, vii, 3, 4, 5, 8, 9, 10, 11, 13, 14, 15, 16, 17, 18, 20, 21, 22, 24, 25, 26, 27, 31, 32, 40, 45, 46, 48, 49, 50, 51, 52, 53, 54,

56, 57, 58, 59, 61, 63, 64, 66, 68, 69, 70, 71, 72, 73, 74, 75, 80, 90, 111, 117, 122, 124, 126, 128, 134, 135, 136, 137, 157,158, 159, 160, 161, 162, 165, 166, 168, 169, 171, 172, 175, 176, 177, 179, 181, 182, 183, 184, 185, 188, 189, 191, 194, 196, 197, 198, 199, 200, 201, 202, 204, 205, 206, 208, 209, 210, 211, 212, 221, 224, 226, 228, 229, 230, 233, 241, 249, 250, 251, 253, 254, 255, 256, 257, 258, 259
bacterial infections, vii, 3, 5, 13, 15, 19, 49, 50, 71, 72, 76, 114, 127, 137, 156, 159, 165, 167, 169, 188, 191, 192, 193, 197, 202, 206, 208, 211, 212, 213, 216, 226, 231, 249, 254, 255, 258
bacterial pathogens, 15, 80
bacterial strains, 59, 198, 204
bacteriophage, 135, 136, 137, 231
bacterium, 162, 165, 166, 167, 176, 182, 194
barriers, 4, 7, 22, 40, 52, 110, 111, 116, 118, 121, 122, 140, 158, 161, 165, 167, 173, 186, 242, 253, 259
basic research, 69, 77, 136, 205, 206, 209, 210, 259
benchmarks, 99, 139
beneficiaries, 24, 96, 97, 141, 143, 145
biotechnology, 80, 86, 90, 198, 203
blood, 51, 142, 185, 189, 216, 219, 225
blood cultures, 142
blood pressure, 189
bloodstream, 16, 49, 250
bronchitis, 19, 105, 129, 252
business model, 87, 89
business partners, 197, 203

C

campaigns, 140, 143, 175, 180, 216
cancer, 69, 70, 105, 157, 160, 169, 177, 178, 189, 241, 252
candidates, 77, 134, 135, 136, 158, 161, 197, 198, 202, 203, 213, 228, 235, 256, 258, 259
Centers for Disease Control and Prevention, vii, 2, 4, 5, 8, 9, 16, 18, 19, 21, 22, 23, 31, 35, 36, 38, 39, 45, 64, 98, 107, 122, 123, 124, 125, 135, 136, 138, 144, 157, 159, 174, 177, 188, 208, 249, 250, 252, 256, 258
challenges, vii, 3, 4, 6, 10, 11, 12, 13, 25, 30, 33, 41, 45, 48, 55, 61, 68, 69, 71, 74, 79, 81, 84, 90, 94, 102, 105, 106, 108, 111, 115, 116, 120, 122, 123, 124, 125, 126, 127, 128, 130, 165, 166, 170, 175, 176, 186, 188, 191, 193, 199, 227, 228, 237
chemical, 8, 11, 23, 68, 123, 196, 199
chemotherapy, 15, 172, 188
children, 21, 32, 99, 104, 105, 129
chronic diseases, 70
clinical judgment, 245
clinical trials, 50, 54, 68, 69, 70, 71, 72, 73, 75, 76, 77, 80, 81, 82, 88, 134, 135, 136, 158, 161, 169, 172, 191, 192, 204, 206, 218, 229, 232, 244, 254, 259
colistin, 61, 157, 159, 160, 162, 171, 176, 179, 182, 183, 184, 197, 200, 205, 206, 208, 210, 213, 218, 219, 221, 222, 250, 253
communication, 9, 54, 190
communities, 30, 163, 175, 178, 180, 223
compounds, 135, 137, 206, 212, 253
contaminated food, 15
contaminated water, 15
cooperative agreements, 50, 77, 78, 101, 140
coordination, 29, 50, 87, 114, 180, 188, 197, 251, 256, 259
cost, 19, 48, 52, 53, 55, 69, 76, 78, 82, 83, 141, 197, 203, 241, 242, 254

covering, 7, 37, 40, 96, 111, 116
crop, 169, 241, 254, 258
culture, 18, 34, 38, 48, 52, 56, 58, 61, 126, 244

D

data analysis, 46
data collection, 38, 44, 46, 99, 102, 106, 109, 116, 118, 139, 180
database, 7, 65, 106, 107, 206, 210
deaths, 9, 16, 41, 42, 43, 44, 66, 74, 177
Department of Agriculture, 27, 209
Department of Defense, 2, 6, 78, 96, 123, 136, 137, 139, 143, 176, 181, 203, 209, 226, 251
Department of Energy, 137
Department of Health and Human Services, 2, 6, 8, 9, 18, 23, 24, 39, 45, 52, 58, 59, 64, 65, 72, 73, 77, 80, 81, 84, 86, 95, 96, 98, 104, 109, 111, 123, 124, 125, 134, 135, 136, 138, 139, 140, 141, 142, 143, 146, 187, 198, 199, 207, 254, 257
detection, 20, 29, 35, 39, 45, 46, 64, 165, 166, 175, 176, 179, 205, 208, 221
diabetes, 157, 160, 189
diagnostic tests, vii, 3, 4, 5, 10, 13, 18, 20, 23, 24, 47, 48, 51, 52, 53, 54, 55, 58, 71, 78, 113, 114, 119, 122, 126, 127, 185, 193, 206, 211, 216, 255
dialysis, 15, 32, 102, 103, 111, 172
disease rate, 34, 37
diseases, 10, 12, 22, 23, 27, 28, 29, 42, 51, 123, 134, 157, 179, 186, 189, 196, 205, 207, 213, 218, 226, 233
doctors, 97, 145, 158, 160, 164, 166, 168, 169, 216, 222, 225, 246, 251, 252
drug discovery, 69, 218
drug industry, viii, 3, 69
drug interaction, 19, 113
drug resistance, 27, 181, 189, 205, 208, 210, 213, 214, 220, 233, 234
drug treatment, 67
drugs, vii, 3, 8, 13, 14, 15, 18, 19, 24, 59, 62, 65, 69, 70, 71, 72, 77, 79, 80, 82, 86, 88, 90, 92, 93, 127, 134, 158, 161, 163, 164, 171, 179, 185, 186, 188, 189, 190, 191, 192, 193, 194, 195, 199, 200, 201, 202, 203, 204, 205, 206, 208, 210, 211, 212, 217, 218, 223, 228, 231, 232, 234, 235, 237, 238, 239, 241, 243, 244, 252, 254, 259

E

E. coli, 157, 159, 164, 166, 171, 176, 182, 183, 184, 188, 200, 202, 250
economic development, 47
economic incentives, 218, 228, 229, 241
economics, 228
education, 20, 98, 138, 140, 163, 220, 221
educational programs, 23
electric current, 231, 232
eligibility criteria, 81, 82, 90, 93, 103
emergency, 19, 95, 141, 142, 158, 160, 247, 251, 252
epidemiologic, 26, 34, 180
epidemiology, 12, 29, 123, 124, 206, 210
evidence, 14, 20, 23, 48, 53, 64, 95, 130, 213, 226, 251
expertise, 30, 48, 96, 124, 130, 143, 189, 198, 203, 245

F

families, 142, 230
FDA approval, 56, 69, 71, 79, 82, 85, 164, 197, 200, 201, 253, 255
federal agency, 13, 33, 48, 127, 128, 145, 199

federal government, vii, 3, 10, 22, 40, 110, 112, 116, 161, 167, 213
financial, 4, 6, 38, 60, 71, 76, 79, 81, 83, 85, 87, 92, 108, 113, 118, 120, 228
financial incentives, 4, 83, 87, 108, 113, 118, 120
financial performance, 81
financial resources, 71
financial support, 60
fiscal year, 38, 40, 50, 77, 81, 82, 83, 86, 135, 136, 142, 143, 159, 161, 171, 175, 178, 206, 209, 222, 223, 256
food, 9, 11, 24, 26, 27, 30, 122, 169, 180, 181, 183, 233, 238, 239, 240, 257
Food and Drug Administration, 2, 11, 18, 24, 27, 54, 57, 58, 59, 60, 61, 62, 63, 64, 65, 67, 70, 72, 73, 75, 77, 79, 80, 86, 123, 135, 174, 180, 187, 199, 209, 248, 254, 255, 257, 258
funding, 24, 38, 39, 48, 49, 50, 52, 53, 76, 77, 78, 85, 90, 91, 92, 108, 127, 134, 137, 158, 161, 163, 171, 175, 178, 181, 196, 197, 201, 203, 204, 206, 209, 211, 212, 223, 224, 236, 237, 244, 256, 259
funds, 23, 28, 71, 77, 78, 134, 135, 136, 206, 210, 212
fungus, 16, 17, 18

G

genes, 15, 56, 181, 223, 250
genome, 29, 175, 179, 181, 183, 206, 209, 210, 211
gonorrhea, 3, 4, 29, 30, 34, 35, 36, 38, 39, 49, 50, 61, 115, 116, 117, 118, 119, 213
grants, 47, 50, 51, 76, 77, 78, 84, 85, 94, 100, 142
growth, 8, 14, 41, 57, 58, 74, 103, 185, 210, 238, 239, 240
guidance, 22, 30, 33, 46, 58, 59, 60, 72, 73, 75, 76, 79, 80, 81, 96, 98, 100, 114, 138, 141, 142, 143, 158, 161, 186, 193, 259
guidelines, 18, 19, 39, 61, 93, 95, 105, 113, 138, 141, 158, 160, 186, 214, 256, 258

H

health, viii, 3, 5, 6, 9, 11, 12, 13, 15, 16, 17, 18, 20, 22, 23, 24, 25, 26, 27, 28, 29, 30, 31, 32, 33, 34, 39, 40, 42, 43, 45, 46, 51, 52, 53, 54, 58, 61, 65, 78, 80, 90, 91, 92, 94, 95, 96, 97, 98, 99, 100, 101, 102, 103, 105, 106, 108, 109, 110, 111, 113, 114, 115, 116, 118, 122, 125, 128, 137, 138, 139, 140, 141, 142, 143, 144, 145, 156, 158, 159, 160, 162, 163, 171, 174, 175, 176, 177, 178, 184, 194, 195, 196, 201, 202, 205, 207, 234, 243, 244, 245, 248, 249, 251, 254, 255, 258
health care, viii, 3, 5, 6, 9, 11, 12, 15, 16, 17, 18, 22, 23, 24, 25, 26, 27, 30, 31, 32, 34, 39, 40, 42, 43, 46, 51, 52, 53, 58, 80, 91, 92, 94, 95, 96, 97, 98, 99, 100, 101, 102, 103, 105, 106, 109, 110, 111, 113, 114, 116, 118, 121, 122, 125, 137, 138, 139, 140, 141, 142, 143, 144, 145, 158, 160, 162, 163, 171, 175, 176, 191, 195, 201, 258
health care costs, 42, 94, 103
health care professionals, 140
health care system, 58, 171
health insurance, 24, 90, 92, 101, 141
health problems, 156, 159, 249
health promotion, 23
human, 8, 9, 11, 15, 20, 23, 24, 26, 27, 29, 68, 79, 122, 134, 175, 176, 179, 180, 183, 184, 189, 204, 221, 239, 244, 246, 250, 256
human health, 9, 11, 15, 23, 122
hydrogen peroxide, 231, 232

hygiene, 20, 22, 162, 172, 233

I

identification, 20, 46, 49, 122, 143, 176, 183, 184, 188, 202, 226, 255
immune defense, 210
immune response, 212
immune system, 31, 74, 169, 212
income, 24, 47, 90, 98, 138, 189
infection, 9, 11, 17, 19, 20, 22, 23, 25, 31, 32, 35, 40, 46, 49, 63, 65, 67, 71, 72, 73, 75, 80, 81, 113, 116, 120, 122, 157, 159, 172, 176, 177, 179, 180, 181, 184, 191, 192, 194, 200, 201, 208, 209, 211, 226, 229, 230, 233, 234, 239, 240, 250, 251, 252
infrastructure, 30, 33, 35, 111, 175, 179, 189, 206, 245
investment, 6, 20, 33, 69, 71, 86, 91, 93, 94, 108, 120, 128, 178, 217, 222, 223, 224, 228, 229, 236, 241, 242, 244, 253
issues, 12, 14, 41, 67, 93, 119, 123, 125, 129, 135, 163, 192, 217, 220, 244, 259

K

kidney, 222, 250
kill, vii, 3, 8, 14, 73, 74, 243

L

laboratory tests, 18
laws and regulations, 27
leadership, 4, 33, 47, 51, 53, 93, 96, 98, 116, 117, 119, 127, 138, 143, 234
learning, 100, 142
legislation, 164, 221, 255
light, 28, 87, 163, 164, 166

M

market economics, 93
market failure, 93
market incentives, 66
marketing, 13, 24, 55, 60, 77, 79, 91, 92, 126, 190
MCR–1, 157, 159, 160, 165, 171, 176, 197, 223, 226
Medicaid, 2, 11, 24, 55, 81, 86, 95, 96, 104, 109, 111, 123, 128, 141, 145, 180, 245, 258
medical, 15, 22, 24, 27, 31, 33, 37, 46, 52, 55, 66, 67, 79, 80, 82, 95, 96, 97, 105, 113, 134, 138, 140, 143, 144, 169, 172, 178, 181, 188, 190, 192, 196, 199, 201, 202, 232, 234, 238, 251, 258
medical care, 46, 181
Medicare, 2, 11, 24, 55, 76, 81, 82, 83, 84, 85, 86, 89, 92, 95, 96, 97, 102, 103, 104, 108, 109, 111, 123, 128, 141, 143, 145, 180, 245, 258
medicine, 97, 145, 157, 160, 172, 177, 178, 188, 189, 198, 200, 208, 226, 238, 239, 249
microorganism, 8
mission, 23, 24, 33, 67, 205, 209
morbidity, 31, 42, 115
mortality, 31, 42, 115, 120
multiple factors, 55, 69, 82, 112, 120

N

National Action Plan, vii, 3, 4, 7, 10, 11, 13, 15, 20, 21, 22, 23, 25, 28, 39, 40, 44, 54, 67, 87, 103, 108, 110, 111, 112, 113, 114, 116, 118, 121, 122, 123, 125, 126, 127, 128, 130, 139, 144, 158, 161, 170, 171, 178, 185, 188, 203, 209, 222, 223, 255, 256, 258

National Institutes of Health, 2, 11, 23, 24, 75, 77, 85, 123, 134, 174, 181, 191, 203, 207, 248, 249, 256
national security, 170, 202
national strategy, 158, 161, 205, 255
nursing, 19, 24, 31, 63, 95, 98, 99, 100, 102, 109, 110, 114, 138, 139, 141, 142, 144, 256
nursing home, 19, 24, 31, 63, 95, 98, 99, 100, 102, 109, 110, 114, 138, 139, 141, 142, 144, 256
nutrients, 57

O

officials, viii, 3, 6, 9, 11, 12, 13, 14, 18, 19, 21, 24, 25, 28, 29, 31, 32, 33, 34, 37, 38, 39, 40, 42, 43, 44, 45, 46, 48, 49, 50, 51, 52, 53, 54, 55, 56, 57, 58, 59, 60, 61, 62, 63, 64, 65, 66, 67, 68, 69, 71, 72, 73, 75, 77, 78, 79, 80, 81, 87, 92, 93, 96, 97,98, 99, 100, 101, 102, 103, 104, 106, 108, 109, 110, 112, 114, 117, 120, 123, 124, 125, 126, 127, 129, 135, 136, 137, 140, 141, 142, 143, 144, 248, 251, 254
opportunities, 62, 100, 109, 142, 244
oral antibiotic, 138
organ, 15, 135, 169, 177, 178, 188, 252
organism, 229, 234
otitis externa, 128
otitis media, 225, 229
outpatient, 18, 19, 21, 96, 97, 98, 99, 100, 103, 106, 110, 111, 138, 139, 141, 142, 143, 144, 163, 182, 193, 214, 216, 252
oversight, 50, 94, 116, 198, 204

P

partial completion, 100, 140
participants, 21, 114, 124, 209
pathogens, 9, 23, 28, 29, 42, 43, 45, 61, 74, 90, 115, 117, 119, 135, 175, 178, 179, 182, 183, 200, 201, 219
patient care, 6, 51, 54, 57, 64, 67, 116, 120
penicillin, 8, 157, 160, 195, 225, 249, 251
pharmaceutical, 69, 71, 85, 88, 89, 90, 101, 141, 161, 189, 203, 205, 208, 217, 228, 254
physicians, 19, 38, 83, 97, 98, 100, 102, 103, 112, 140, 142, 145, 162, 163, 193, 194, 214, 246
pipeline, 4, 6, 9, 68, 69, 71, 73, 86, 87, 92, 94, 116, 120, 172, 185, 196, 197, 201, 204, 211, 216, 217, 228, 241, 242, 254
policy, 13, 44, 75, 82, 83, 87, 89, 93, 96, 123, 128, 143, 144, 255
policy makers, 90, 93
policymakers, 45, 115
polymerase chain reaction, 183
population, 3, 5, 6, 26, 30, 34, 37, 38, 39, 46, 48, 73, 80, 91, 109, 115, 117, 119, 164, 168, 179, 186, 189, 192, 219, 243, 255
portfolio, 78, 197, 201, 202, 205, 207, 209, 217, 224, 228
prevention, 11, 20, 22, 23, 31, 45, 100, 122, 142, 175, 176, 178, 179, 181, 186, 190, 223, 226, 233, 246, 256
principles, 61, 70, 93, 128, 245
private practice, 37, 246
private sector, 188, 200
procurement, 54, 196, 199, 203, 244
professionals, 46, 157, 159, 169, 249, 258
progress reports, 7, 22, 40, 111, 116, 121
project, 35, 38, 39, 50, 78, 91, 98, 128, 136, 142, 203, 210
public health, 8, 11, 23, 24, 25, 26, 27, 28, 32, 33, 34, 39, 45, 46, 64, 65, 70, 90, 101, 111, 114, 115, 117, 118, 120, 123, 124, 125, 140, 156, 159, 161, 163, 164, 165, 166, 167, 168, 170, 173, 176, 178, 179, 182, 183, 184, 187, 188, 192, 194,

196, 197, 201, 202, 209, 227, 243, 244, 248, 258
public policy, 68
public service, 17
public-private partnerships, 23, 76, 77, 172, 196, 201, 203, 243

R

reallocation of resources, 197, 203
recognition, 60, 61, 62, 83, 140, 181
recommendations, iv, 3, 4, 22, 34, 40, 41, 62, 87, 90, 93, 111, 117, 118, 121, 124, 137, 138, 143, 176, 218, 236, 255, 256
regulatory bodies, 48
regulatory framework, 194
regulatory requirements, 48, 71
rehabilitation, 32, 95, 99, 141
reliability, 12, 14, 125, 129
representativeness, 3, 5, 34, 35, 37, 39, 45, 115, 117, 119
resistant infections, vii, viii, 3, 4, 6, 9, 13, 15, 20, 21, 29, 31, 41, 42, 47, 48, 68, 74, 111, 121, 123, 127, 131, 136, 157, 159, 178, 180, 181, 182, 188, 193, 196, 197, 211, 243, 249, 250, 255
resources, 22, 23, 38, 40, 45, 48, 53, 56, 62, 83, 93, 97, 98, 109, 110, 111, 116, 118, 119, 138, 140, 142, 143, 144, 175, 178, 190, 195, 198, 203, 204, 206, 210, 217, 220, 223
response, 3, 28, 29, 35, 39, 45, 75, 79, 95, 112, 116, 118, 119, 121, 145, 156, 157, 160, 164, 165, 166, 167, 176, 179, 180, 182, 184, 201, 212, 214, 226, 235, 236, 240, 246, 251, 255, 259
retail, 19, 24, 27, 103, 110, 158, 160, 175, 183, 252
risk, 19, 45, 56, 65, 66, 67, 94, 109, 134, 157, 160, 162, 171, 172, 181, 192, 197, 203, 206, 210, 219, 230, 257

S

safety, 19, 24, 28, 63, 66, 67, 68, 70, 71, 73, 75, 95, 103, 137, 138, 141, 190, 199, 240
science, 135, 137, 165, 167, 191, 218, 228
scope, 13, 14, 21, 28, 66, 93, 109, 110, 126
sequencing, 29, 175, 179, 181, 197, 206, 209, 210, 211
services, iv, 38, 77, 82, 97, 103, 134, 144, 145, 210, 245
superbug, 157, 159, 160, 162, 164, 165, 166, 167, 240, 249, 250, 251
surveillance, vii, 3, 4, 10, 12, 15, 20, 25, 26, 27, 28, 29, 30, 31, 34, 35, 36, 37, 39, 40, 42, 44, 45, 46, 47, 48, 64, 67, 99, 108, 111, 114, 115, 117, 119, 122, 124, 125, 139, 170, 172, 179, 183, 206, 208, 210, 226, 240, 255, 256, 257
susceptibility, 13, 27, 29, 37, 38, 41, 57, 62, 126, 143, 183, 185, 186, 194, 197
symptoms, 29, 113

T

target, 21, 24, 68, 72, 75, 78, 105, 111, 114, 135, 136, 140, 180, 192, 201, 209, 210, 211, 212, 215
technical assistance, 98, 100, 142
technical comments, 118
technical support, 118
testing, vii, viii, 3, 4, 5, 10, 12, 18, 26, 28, 29, 30, 35, 36, 38, 40, 46, 47, 49, 50, 51, 56, 58, 77, 117, 124, 129, 179, 180, 183, 191, 194, 206, 212, 216, 257
therapeutics, 20, 105, 191, 206, 209, 210, 212, 228, 232, 255, 256
therapy, 35, 61, 80, 138, 140, 188, 193, 197, 213
threats, 8, 13, 16, 17, 23, 28, 30, 34, 45, 48, 49, 64, 65, 114, 115, 120, 125, 169, 174,

176, 178, 179, 182, 184, 196, 197, 198, 199, 202, 204, 208, 245, 256, 258, 259
tracks, 5, 25, 27, 29, 31, 32, 38, 68, 99, 139, 183, 254
training, 20, 46, 96, 100, 102, 108, 112, 113, 114, 140, 141, 142, 143, 144, 181
treatment, vii, 3, 9, 10, 15, 19, 24, 34, 35, 41, 51, 52, 55, 56, 58, 61, 70, 71, 72, 73, 74, 79, 80, 82, 94, 96, 98, 103, 104, 105, 116, 120, 129, 138, 140, 142, 143, 157, 160, 162, 165, 166, 169, 188, 189, 190, 191, 192, 194, 197, 202, 211, 212, 213, 249, 251, 256
trial, 72, 73, 76, 134, 135, 188, 189, 190, 191, 192, 206, 211, 213, 218

U

U.S. Department of Agriculture, 27, 135, 180, 257
United States, v, 1, 2, 4, 5, 7, 8, 9, 12, 13, 14, 15, 17, 18, 19, 20, 23, 24, 25, 28, 30, 32, 34, 35, 38, 42, 43, 44, 48, 52, 57, 60, 61, 74, 81, 86, 87, 95, 96, 99, 103, 106, 107, 115, 119, 122, 123, 124, 126, 130, 134, 139, 143, 144, 145, 157, 158, 159, 160, 166, 169, 171, 172, 174, 176, 177, 184, 188, 208, 209, 210, 215, 220, 222, 233, 234, 244, 246, 250, 251, 252, 253, 257, 258, 259
updating, 61, 62, 63, 66, 67, 116, 118, 120, 194, 195
upper respiratory infection, 104, 105, 129
urinary tract, 72, 162, 191, 200, 202, 250
urinary tract infection, 72, 162, 191, 200, 202, 250

V

venture capital, 188, 236
victims, 177, 178, 252
viral infection, 49, 113, 193, 212, 213, 219, 223
virus infection, 216
viruses, 19, 74, 112, 167, 199, 244

W

witnesses, 159, 161, 162, 163, 164, 170, 172, 173, 174, 227, 230, 247
World Health Organization, 9, 10, 20, 22, 24, 47, 52, 106, 123, 133, 135, 143, 157, 159, 220, 234, 249, 253
worldwide, 24, 52, 71, 135, 198, 204, 243